SpringerBriefs in Mathematical Physics

Volume 34

T0254611

SpringerBriefs are characterized in general by their size (50–125 pages) and fast production time (2–3 months compared to 6 months for a monograph).
Briefs are available in print but are intended as a primarily electronic publication to be included in Springer's e-book package.

Typical works might include:

- An extended survey of a field
- A link between new research papers published in journal articles
- A presentation of core concepts that doctoral students must understand in order to make independent contributions
- Lecture notes making a specialist topic accessible for non-specialist readers.

SpringerBriefs in Mathematical Physics showcase, in a compact format, topics of current relevance in the field of mathematical physics. Published titles will encompass all areas of theoretical and mathematical physics. This series is intended for mathematicians, physicists, and other scientists, as well as doctoral students in related areas.

More information about this series at http://www.springer.com/series/11953

Stefan Hollands · Ko Sanders

Entanglement Measures and Their Properties in Quantum Field Theory

 Springer

Stefan Hollands
Institute for Theoretical Physics
University of Leipzig
Leipzig, Sachsen, Germany

Ko Sanders
School of Mathematical Sciences
Dublin City University
Dublin, Ireland

ISSN 2197-1757 ISSN 2197-1765 (electronic)
SpringerBriefs in Mathematical Physics
ISBN 978-3-319-94901-7 ISBN 978-3-319-94902-4 (eBook)
https://doi.org/10.1007/978-3-319-94902-4

Library of Congress Control Number: 2018951916

This Springer imprint is published by the registered company Springer Nature Switzerland AG
The registered company address is: Gewerbestrasse 11, 6330 Cham, Switzerland

Acknowledgements

S. H. likes to thank J. Eisert, P. Grangier and R. Longo for discussions, and O. Islam for help with figures. K. S. would like to thank Leipzig University, where this project was carried out, and U. Rome II ("Tor Vergara") for funding a visit in November 2016 and for the opportunity to present some of the results in this volume in a seminar. He also gratefully acknowledges an invitation to MFO where some of the results relevant to this volume were presented during the workshop "Recent mathematical developments in quantum field theory" (ID1630). We thank Y. Tanimoto for pointing out an error concerning the ordering of entanglement measures in an earlier version of this volume and K.-H. Rehren for comments on our manuscript.

Contents

Abstract

An entanglement measure for a bipartite quantum system is a state functional that vanishes on separable states and that does not increase under separable (local) operations. For pure states, essentially all entanglement measures are equal to the v. Neumann entropy of the reduced state, but for mixed states this uniqueness is lost. In quantum field theory, bipartite systems are associated with causally disjoint regions. But if these regions touch each other, there are no separable normal states to begin with, and one must hence leave a finite "safety corridor" between the regions. Due to this corridor, the normal states of bipartite systems are necessarily mixed, so the v. Neumann entropy is not a good entanglement measure anymore in this sense. In this volume, we study various good entanglement measures. In particular, we study the relative entanglement entropy, E_R, defined as the minimum relative entropy between the given state and an arbitrary separable state. We establish upper and lower bounds on this quantity in several situations: (1) In arbitrary CFTs in $d+1$ dimensions, we provide an upper bound for the entanglement measure of the vacuum state if the two regions of the bipartite system are a diamond and the complement of another diamond. The bound is given in terms of the spins, dimensions of the CFT, and the geometric invariants associated with the regions. (2) In integrable models in $1+1$ dimensions defined by a general analytic, crossing symmetric two-body scattering matrix, we give an upper bound for the entanglement measure of the vacuum state for a pair of diamonds that are far apart, showing exponential decay with the distance between the diamonds. The class of models includes, e.g., the Sinh–Gordon field theory. (3) We give upper bounds for our entanglement measure for a free Klein–Gordon/Dirac field in the ground state on an arbitrary static spacetime. Our upper bounds show exponential decay of the entanglement measure for large geodesic distance and an "area law" for small distances (modified by a logarithm). (4) We show that if we add charged particles to an arbitrary state, then E_R decreases by a positive amount which is no more than the logarithm of the quantum dimension of the charges (this dimension need not be an integer). (5) We establish a lower bound on our entanglement measure for arbitrary regions that get close to each other. This lower bound is of the type of an "area law" with the proportionality constant given by the number N of free fields in the UV fixed point times a quantity D_2 that can be interpreted as the distillable entanglement of one "Cbit pair" in the state.

Chapter 1
Introduction

Abstract Entanglement measures quantify the amount of entanglement between parts of a system, but a considerable part of the literature in Quantum Information Theory has focussed on quantum systems with finitely many degrees of freedom. In this volume, we will focus on the question whether qualitatively new features can arise due to the presence of infinitely many degrees of freedom.

While correlations between different parts of a system can exist both in classical and quantum physics, there can exist in quantum systems certain more subtle correlations that are absent in classical ones. Such correlations are nowadays referred to as the "entanglement" between the subsystems. Historically, the first quantitative measure of entanglement were the Bell-inequalities [1, 2]—or rather, their violation.

Motivated not least by technological advances in controlling and manipulating quantum systems, there has by now emerged an understanding of certain types of operations that one can think of, in a definite way, as not increasing the entanglement originally present in a bi-partite quantum system (see e.g. [3] for a review). The set of these operations, often called "LOCC-operations"[1]—as well as various "asymptotic" generalizations thereof, where one is allowed to access and manipulate arbitrarily many copies of the given bipartite system—give the set of states on a bipartite quantum system an ordering: A state σ_1 is not more entangled than a state σ_2, if σ_1 can be obtained from σ_2 by LOCCs.

On the one extreme, one has states that are not entangled at all. These are called "separable" and are described by statistical operators σ of the form

$$\sigma = p_1 \rho_{A1} \otimes \rho_{B1} + p_2 \rho_{A2} \otimes \rho_{B2} + \dots , \tag{1.1}$$

---

[1]This stands for "local operations and classical communications". In this volume, we will actually use an even broader class.

© The Author(s), under exclusive licence to Springer Nature Switzerland AG 2018
S. Hollands and K. Sanders, *Entanglement Measures and Their Properties*
in Quantum Field Theory, SpringerBriefs in Mathematical Physics 34,
https://doi.org/10.1007/978-3-319-94902-4_1

where, according to the usual principles of quantum theory, the total Hilbert space is the tensor product $\mathcal{H} = \mathcal{H}_A \otimes \mathcal{H}_B$, where ρ_{A1}, ρ_{A2} etc. respectively ρ_{B1}, ρ_{B2} etc. are statistical operators for subsystem A respectively B, and where $p_i \geq 0, \sum_i p_i = 1$. In classical physics, all states are separable.[2] On the other hand, in quantum physics, one has non-separable, i.e. entangled, states. In particular, one has maximally entangled states. In between these extremes, one has states that can neither be manipulated using LOCCs into a maximally entangled state, nor be obtained from separable states by such operations. In general, the ordering is only partial: we cannot say for each and every pair of states whether one or the other is more entangled.

To understand better the structure of entangled states, it is useful to introduce entanglement measures. At a bare minimum,[3] an entanglement measure $E(\rho)$ should clearly satisfy the following properties:

(I) $E(\rho)$ should give a number in $[0, \infty]$, returning 0 for separable states,

and

(II) $E(\rho)$ should be monotonically decreasing under LOCCs.

A great variety of entanglement measures has been introduced in the literature. While a classification seems presently out of reach, a particularly simple and satisfactory story emerges for the set of pure states $\rho = |\Psi\rangle\langle\Psi|$. Here, one can show [4] under moderate and reasonable technical assumptions, that every entanglement measure is, up to a change of normalization, equal to the v. Neumann entropy of the reduced density matrix for subsystem A (or equivalently B), i.e. $E(\rho) = H_{\text{vN}}(\rho_B) = H_{\text{vN}}(\rho_A)$. As usual, the reduced density matrix $\rho_A = \text{Tr}_B\rho$ corresponds to the restriction to subsystem A, and similarly for B. Furthermore, it can be shown [5] that if two pure states have the same $E(\rho)$, then they can be converted into each other "asymptotically" by LOCC operations.

Unfortunately, for general mixed states ρ, uniqueness is lost and one can say under the same types of technical assumptions only that a general entanglement measure E must always yield values between two extreme, conceptually distinguished entanglement measures called "entanglement cost" and "distillable entanglement". Furthermore, for mixed states, the v. Neumann entropy of the reduced density matrix does not provide a reasonable measure, as it can for instance return the same value for separable and maximally entangled states.

A considerable part of the literature in Quantum Information Theory has focussed on quantum systems with finitely many degrees of freedom—leading at the technical level mostly to (complicated) questions about algebras of matrices—and, furthermore, to some extent, on an underlying assumption that the kinematics is non-

---

[2]In classical physics, if μ is a measure on a product phase space $X = X_A \times X_B$ which is, say, absolutely continuous relative to the Lebesgue measure, then we can approximate it with arbitrary precision by sums of product measures $\sum_i \mu_{Ai} \times \mu_{Bi}$ (e.g. on the dense subspace of smooth observables).

[3]It may or may not be possible/desirable to also have other properties such as convexity under convex linear combinations.

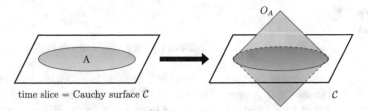

Fig. 1.1 Causal diamond associated with A

relativistic. It is therefore interesting to extend the analysis to relativistic quantum field theory (QFT) where both assumptions no longer hold. One may ask:

(QI) Whether a relativistic setup will lead to modified concepts of classical communication, distillation protocols, etc. at a kinematical level.

(QII) Whether qualitatively new features can arise due to the presence of infinitely many degrees of freedom.

In this volume, we will basically ignore (QI) and work with essentially the same concepts of LOCCs as in the standard theory; see e.g. [6–8] and references therein for further developments in direction (QI). Instead, we will focus on (QII). First of all, we note that, in QFT, the notion of subsystem is always tied to the localization in spacetime. Thus, if A is some subset of a Cauchy surface (i.e., a "time slice") in Minkowski spacetime (or, more generally, a globally hyperbolic curved Lorentzian spacetime), then one ascribes [9, 10] to A a set of observables $\mathfrak{A}_A$ localized in the "causal diamond" O_A with base A, see Fig. 1.1.

Informally speaking, $\mathfrak{A}_A$ is the algebra generated by the quantum fields localized at points in O_A. More precisely, $\mathfrak{A}_A$ is the v. Neumann algebra generated by the spectral projections of the quantum fields that are "smeared" against a test function supported in O_A. It is a standard property of relativistic quantum field theories that if a region B on the same time-slice as A is disjoint from A—so that there is no causal curve connecting O_A with O_B—then

$$[\mathfrak{A}_A, \mathfrak{A}_B] = \{0\} . \tag{1.2}$$

This relation of course also holds for non-relativistic quantum mechanical systems, since the algebra of observables $\mathfrak{A} = \mathfrak{A}_A \vee \mathfrak{A}_B$ generated by $\mathfrak{A}_A$ and $\mathfrak{A}_B$ is by definition set up in the form of a tensor product $\mathfrak{A}_A \otimes \mathfrak{A}_B$ where the factors act on $\mathcal{H}_A \otimes \mathcal{H}_B$.

At this point, however, an—at first sight seemingly academic—difference arises in QFT. This difference has its origins in a mathematical fact about v. Neumann algebras. As is well-known, there are different "types" of v. Neumann algebras [11, 12]. The algebras appearing in non-relativistic systems, for instance matrix algebras, are typically of so-called type I, whereas the algebras appearing in QFT are typically of type III [13, 14]. It is not so important for us what precisely these types mean (see e.g. [15, 16] for general references). The key point for us is rather that for type III, unlike type I,

$$[\mathfrak{A}_A, \mathfrak{A}_B] = \{0\} \quad \text{does } not \text{ always imply } \quad \mathfrak{A} \cong \mathfrak{A}_A \otimes \mathfrak{A}_B \,, \tag{1.3}$$

where "$\cong$" means "up to unitary equivalence". In fact, the conclusion—which is closely related to the "split property" (see [13, 17, 18] and below)—only holds if A and B are separated by a *finite* corridor, but it does not hold for instance when B is the complement of A. Thus, it fails in the usual and most natural situation wherein we partition the total system into the union of two disjoint subsystems. For further discussion on this and related issues see [19] and the nice review of [20], which is directed at a wider theoretical physics audience.

This seemingly academic difference between QFT and quantum systems with finitely many degrees of freedom has the following important consequence. If $\mathfrak{A} \cong \mathfrak{A}_A \otimes \mathfrak{A}_B$ holds—in which case we say that $\mathfrak{A}_A$ and $\mathfrak{A}_B$ are "statistically independent" [21]—then we can define separable states as in (1.1) for the total system. On the other hand, if $\mathfrak{A}_A$ and $\mathfrak{A}_B$ only commute but are not statistically independent, then we cannot have (normal) separable states. Thus, we are entirely outside the usual setup for discussing entanglement in Quantum Information Theory. In particular, we are outside this framework if B is the complement of A on a time slice. On the other hand, if we leave a finite safety corridor between A and B, then $\mathfrak{A}_A$ and $\mathfrak{A}_B$ are typically statistically independent, and the usual concepts from Quantum Information Theory such as separable states, LOCCs, etc. carry over.

Thus, in QFT, we *should leave a finite safety corridor between A and B*. But then it is clear that if we start with a state of the full quantum field theory, ρ (for instance the vacuum state $\rho_0 = |0\rangle\langle 0|$), then, since $A \cup B$ has an open complement C (the corridor), the restriction of ρ to $\mathfrak{A}_A \otimes \mathfrak{A}_B$ is never a pure state, as we shall prove rigorously below in Sect. 2.3. Therefore, following our general discussion about entanglement measures, we no longer have a unique measure with which to quantify the entanglement of a state ρ across A and B. In particular, the v. Neumann entropy does *not* yield a satisfactory entanglement measure.

We are thus forced in relativistic QFT to consider alternative entanglement measures with good properties [at least (I) and (II)] for mixed states. In this volume, we shall study several such measures in the framework of algebraic QFT [9, 10]. The measure which we shall focus on mostly is the so-called "relative entropy of entanglement" proposed in [5]. This measure is based on Umegaki's relative entropy functional [22] $H(\rho, \sigma) = \text{Tr}(\rho \ln \rho - \rho \ln \sigma)$, or rather its generalization to v. Neumann algebras of general type due to Araki [23, 24]. The relative entanglement entropy is given by

$$E_R(\rho) = \inf_{\sigma \text{ separable}} H(\rho, \sigma) \,. \tag{1.4}$$

$E_R(\rho)$ may be interpreted as the information that we can expect to gain if we update our belief about the state of the system from being separable to ρ [25]. Due to the variational definition [infimum over all separable states, cf. (1.1)], it is not even clear, a priori, whether $E_R(\rho) > 0$, nor is it clear that $E_R(\rho) < \infty$ for any state in the QFT

Fig. 1.2 The regions A and B

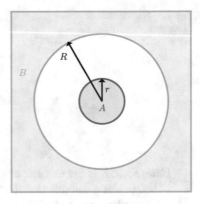

setting.[4] Our aim is thus to investigate E_R and provide upper and lower bounds in several situations.

1.1 Summary of Main Results

Our main results can be summarized as follows:

1. Let A be a ball of radius r in a $t = 0$ time slice of Minkowski space and B the complement of a concentric ball with radius $R > r$ (see Fig. 1.2). In any conformal field theory in $d+1$ dimensions, we have

$$E_R(\rho_0) \leq \ln \sum_{\mathcal{O}} \left(\frac{r}{R}\right)^{d_{\mathcal{O}}}, \tag{1.5}$$

where $\rho_0 = |0\rangle\langle 0|$ is the vacuum state,[5] and where $d_{\mathcal{O}}$ are the dimensions of the local operators $\mathcal{O}$ in the theory.

If, more generally, the diamonds are not necessarily concentric, we can introduce the conformally invariant cross ratio $u = \frac{(x_{B+}-x_{B-})^2(x_{A+}-x_{A-})^2}{(x_{A-}-x_{B-})^2(x_{A+}-x_{B+})^2}$ and similarly v, see (4.120), associated with upper/lower tip of diamond O_A and the upper lower tip of diamond O'_B, see Fig. 1.3. If $\tau, \theta \in \mathbb{R}$ are the functions of these cross ratios u, v defined below in Eq. (4.121), then we get in 3+1 dimensions

$$E_R(\rho_0) \leq \ln \sum_{\mathcal{O}} e^{-\tau d_{\mathcal{O}}} \left[2S_{\mathcal{O}}^R + 1\right]_{\theta} \left[2S_{\mathcal{O}}^L + 1\right]_{\theta}. \tag{1.6}$$

---

[4]In fact, as shown in [26], entanglement measures that are well-behaved in the type I-setting can become ill-defined for type III, as is the case e.g. for the "entanglement of formation". [26] has also shown that the entanglement entropy $E_R(\rho_0)$ behaves well under a "nuclearity condition", a technique to which we will come back in the body of the volume.

[5]In the body of this volume we will distinguish, for technical reasons, the expectation value function of a statistical operator $\omega(\,.\,) = \text{Tr}(\,.\,\rho)$ and the statistical operator ρ itself.

Fig. 1.3 Nested causal
diamonds

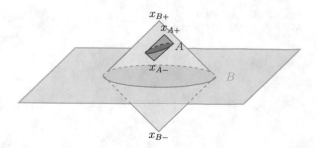

Here, $S_{\mathcal{O}}^L$, $S_{\mathcal{O}}^R$ is the number of primed/unprimed spinor indices of the operator $\mathcal{O}$ and $[n]_\theta$ is a suitably defined "quantum deformed" natural number n. A similar bound is obtained also for chiral conformal field theories. When the outer diamond is much larger, $r/R \ll 1$, our bound gives for instance

$$E_R(\rho_0) \lesssim N_{\mathcal{O}} \left(\frac{r}{R}\right)^{d_{\mathcal{O}}} \tag{1.7}$$

for concentric diamonds, where $\mathcal{O}$ is the operator with the smallest non-trivial dimension $d_{\mathcal{O}}$, and $N_{\mathcal{O}}$ the multiplicity of such operators. This result is consistent with the "small x expansion" obtained by [27, 28] in 1+1 dimensions.

2. Let ω be any state of finite energy, and let $\chi^*\omega$ be a state obtained by adding "charges" $\chi = \prod \chi_i^{n_i}$ (in a generalized sense which may include charged pseudo-particles with braid-group statistics [29–31]) in region A or B as indicated in Fig. 1.4. We have

$$0 \le E_R(\omega) - E_R(\chi^*\omega) \le \ln \prod_i \dim(\chi_i)^{2n_i}, \tag{1.8}$$

irrespective of the nature of ω, or the relative position of A and B, or the dimension $d+1$ of spacetime. Here, n_i is the number of irreducible charges χ_i of type i, and $\dim(\chi_i)$ the "quantum dimension" of the charge. For instance, this dimension is N if the charge is created by a local operator transforming in the fundamental representation of $O(N)$, but can also be non-integer e.g. for anyonic pseudo-particles in 1+1 dimensions.

Fig. 1.4 Charges χ_i added
to A

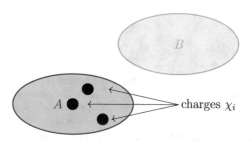

3. For the real Klein-Gordon scalar QFT with field equation

$$\Box\phi - m^2\phi = 0 \,, \tag{1.9}$$

and positive mass m on an arbitrary ultra-static, globally hyperbolic spacetime $\mathcal{M}$ of dimension d+1 with metric $-dt^2 + h_{ij}(x)dx^i dx^j$, we show that the entanglement of the ground state ρ_0 of the theory decays for $mr \gg 1$ as[6]

$$E_R(\rho_0) \lesssim C_\infty\, e^{-mr/2} \,, \tag{1.10}$$

where r is the geodesic distance (with respect to h_{ij}) between A and B in a static slice, see Fig. 1.5 and where C_∞ is some constant.

For the Majorana Dirac QFT with field equation

$$(\nabla\!\!\!/ + m)\psi = 0 \,, \tag{1.11}$$

and non-vanishing mass, we also have an upper bound when the geodesic distance r between A and B goes to zero in a static slice, in a spacetime of the product form $\mathcal{M} = \mathbb{R}^{1,1} \times \Sigma$. More precisely, let $A = \{(t, x, y) \mid x < 0, t = 0\}$ and $B = \{(t, x, y) \mid x > r, t = 0\}$ (where (t, x) are standard coordinates on $\mathbb{R}^{1,1}$ and $y \in \Sigma$), and let ρ_0 be the ground state. Then as $mr \to 0$, we have the upper bound

$$E_R(\rho_0) \le C_0|\ln(mr)| \sum_{j\le d-1} r^{-j} \int_{\partial A} a_j \lesssim c_0|\ln(mr)|\frac{|\partial A|}{r^{d-1}} \,, \tag{1.12}$$

where we assume $d \ge 2$ and the a_j are geometric invariants associated with a heat kernel on $\partial A \cong \Sigma$ and C_0, c_0 are constants. We expect our methods to yield similar results for general spacetimes with bifurcate Killing horizon, Fig. 1.6, see [32] for this notion.

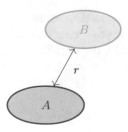

Fig. 1.5 The regions A and B

---

[6]Formula (1.13) below suggests that the upper bound can be improved to $C_\infty(\delta)e^{-mr(1-\delta)}$ for each $\delta > 0$.

Fig. 1.6 Spacetime with
bifurcate Killing horizon

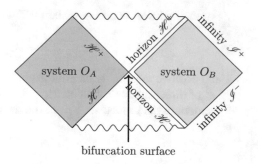

bifurcation surface

4. For the class of massive integrable models on 1+1 dimensional Minkowski space
   with factorizable two-body scattering matrix S_2 of the general form (4.64), and
   A and B given by two half-lines of the time slice $\mathbb{R}$ separated by $r > 0$, the
   vacuum state $\rho_0 = |0\rangle\langle 0|$ satisfies the following bound. For $mr \gg 1$ (here m
   is the mass) we have for any $\kappa > 0$ such that $S_2(\zeta)$ has no poles in the strip
   $\{\zeta \in \mathbb{C} : -\kappa < \Im\zeta < \pi + \kappa\}$, and any small[7] $\delta > 0$,

$$E_R(\rho_0) \lesssim \frac{4e}{\kappa} \sqrt{\frac{\|S_2\|_\kappa}{\pi mr}} \, e^{-mr(1-\delta)} . \qquad (1.13)$$

where $\|S_2\|_\kappa$ is the supremum of $|S_2(\zeta)|$ in the strip. The constant in the bound
diverges when the poles of S_2 approach the "physical strip". Since poles inside
the physical strip are characteristic for models with bound states, we can say that
our upper bound on the entanglement entropy deteriorates as we approach this
situation.

Our bound applies in particular to the *Sinh-Gordon model* with equation of
motion

$$\Box\phi - m^2\phi - g^2 \sinh \phi = 0, \qquad (1.14)$$

where $S_2(\theta) = \frac{\sinh\theta - i\sin b}{\sinh\theta + i\sin b}$, $b = \frac{\pi g^2}{1+g^2}$, and the constants C are given in terms of
b or g. In this case, the constants deteriorate for $g \to \infty$.

5. Consider a massive QFT on d+1 dimensional Minkowski space satisfying a
   "nuclearity condition" in the sense of [33], and let A and B regions separated by
   $r \gg$ than the size of A and B, see Fig. 1.5.
   We show that the entanglement in the vacuum state ρ_0 has sub-exponential decay,
   i.e.

$$E_R(\rho_0) \lesssim C \, e^{-(mr)^k} , \qquad (1.15)$$

---

[7] We cannot put κ or δ to zero, since the asymptotic bound holds, roughly speaking, when $1/(\delta\kappa) \lesssim mr$.

for any given $k < 1$ (our C diverges when $k \to 1$). Since such nuclearity conditions have been shown for massive free fields of spin 0, 1/2 [18, 34], the bound holds in particular for such theories.

6. Consider a QFT in d+1 dimensions satisfying a suitable "nuclearity condition" for a thermal state ρ_β, cf. (4.4), and let A and B again be regions separated by $r \gg$ than the size of A and B, see Fig. 1.5. We show:

$$E_R(\rho_\beta) \lesssim C \, r^{-\alpha+1} \,, \tag{1.16}$$

where α is a parameter entering the nuclearity condition.

7. For any conformal QFT in $d+1$ dimensions with vacuum state $\rho_0 = |0\rangle\langle 0|$, and for A and B regions separated by a thin corridor of diameter $\epsilon > 0$, we show that asymptotically, as $\epsilon \to 0$

$$E_R(\rho_0) \gtrsim \begin{cases} D_2 \cdot |\partial A| / \epsilon^{d-1} & d > 1, \\ D_2 \cdot \ln \frac{\min(|A|,|B|)}{\epsilon} & d = 1, \end{cases} \tag{1.17}$$

("area law"), where D_2 is the distillable entropy of an elementary "Cbit" pair (defined in Sect. 5.1). As we point out in the text, the same argument shows that for a non-conformal, asymptotically free theory and for any state ω with finite energy in $d > 1$ spatial dimensions, one would get the bound

$$E_R(\omega) \gtrsim N \cdot D_2 \cdot |\partial A| / \epsilon^{d-1} \tag{1.18}$$

where N is the number of independent free fields in the short-distance scaling limit, for instance $N = n^2 - 1$ in SU(n) pure Yang-Mills theory.

For a massive Dirac field, we have found a qualitatively similar upper bound in (3), so these lower bounds should be expected to be qualitatively sharp.

In order to obtain these results, we use several other entanglement measures which give upper and lower bounds on E_R and which are often easier to estimate, such as E_D (distillable entropy), E_N (logarithmic dominance), E_M (modular entanglement), and others. Some of these are of independent interest, and to our knowledge, new. A table comparing these entanglement measures is presented in Sect. 3.8. A key role is also played in our proofs by techniques from Tomita-Takesaki modular theory for v. Neumann algebras and their—to a large extent well-known—relation to quantum field theory. These come in on the one hand via their relation with spacetime symmetries (Bisognano-Wichmann [35] and Hislop-Longo [36] theorems) and on the other hand via their relationship with "nuclearity bounds" [18, 33, 37, 38], both of which are combined with methods from complex analysis. We note that the usefulness of nuclearity bounds in the context of entanglement has been appreciated already by [26], and they have been used also more recently by [39], which appeared after our work was completed. Among the other tools that we use is the theory of superselection sectors [29–31, 40–42].

1.2   Comparison with Other Approaches to Entanglement in QFT

It is worth commenting on the conceptual and practical differences between our approach and the substantial body of literature on entanglement in QFT based on the "replica trick", or "holographic" methods, see e.g. [43, 44] for reviews with many references.

In the "replica trick" [45], which applies most straightforwardly to ground states on static spacetimes, one ignores the problems discussed above with the v. Neumann entropy, and formally represents $H_n(\omega_A) = \frac{1}{1-n}(\ln Z(\mathcal{M}_n) - n \ln Z(\mathcal{M}_1))$. Here, $\omega_A = \mathrm{Tr}_A |0\rangle \langle 0|$ is the reduced state, H_n is a regulated version of the v. Neumann entropy,[8] and $\mathcal{M}_n$ is an n-sheeted cover of $\mathcal{M}$ obtained by gluing n copies of $\mathcal{M}$ across the boundary of A. Z is the the partition function of the corresponding Euclidean QFT, often represented in terms of a functional integral, or "defect operators" as in the pioneering paper [46] on 1+1 dimensional conformal field theories. In either case, the result is divergent due to the conical singularity on $\mathcal{M}_n$ along ∂A, but one can at this stage introduce a short distance (UV) cutoff ϵ of some sort, and get a finite answer, $H_{vN}^{\epsilon}(\omega_A)$. The divergent terms are often found to be organized in a series in inverse powers of ϵ. The most divergent term is usually $\propto |\partial A|/\epsilon^{d-1}$, $d > 1$ ("area law" [47–49]), and the sub-leading ones are often—though not always [50]—given in terms of curvature invariants associated with ∂A.

Compared with our entanglement measure (1.4), one is tempted to perhaps expect a relationship of the form

$$H_{vN}^{\epsilon}(\omega_A) \sim E_R(\omega) \qquad (1.19)$$

when $\epsilon \sim \mathrm{dist}(A, B)$ becomes small compared to the volume of A and all other length scales in the QFT (including any scales introduced by the state $\omega = |\Psi\rangle\langle\Psi|$ if it is not the vacuum), and when B approaches the complement of A. Some of our results indicate that the above relation may indeed be roughly correct in many cases, but (3) and (4) indicate that this is perhaps so only in massive theories, and perhaps only up to powers in $\ln(m\epsilon)$.

Compared to our approach, the replica trick has, at any rate, a rather different distribution of strengths and weaknesses. The strengths are that the basic formulas are, although formal, strikingly elegant, and in principle concrete, making rather non-trivial computations possible in many interesting examples, and establishing also an interesting link to other ideas in quantum field theory, such as e.g. the c-theorem [51, 52].

The weakness is that, in order to obtain a finite answer in the limit $\epsilon \to 0$, one must either subtract by hand the divergent terms, or consider differences of quantities like for instance $H_n(\mathrm{Tr}_A|0\rangle\langle 0|) - H_n(\mathrm{Tr}_A|\Psi\rangle\langle\Psi|)$, where $|\Psi\rangle$ is some reference state, or like the "mutual information" for disjoint regions A, B—hoping that the divergences are state-independent and cancel (this is not always the case [50]).

---

[8]It is defined as $H_n(\rho) = \frac{1}{1-n} \ln \mathrm{Tr}\rho^n$.

At any rate, one does *not* obtain a quantity that satisfies the basic postulates (I) and (II), and furthermore, the assumption of a short distance cutoff at intermediate stages invalidates the basic assumptions of locality (if the cutoff is imposed in momentum space), or relativistic covariance (if the cutoff is implemented on a lattice), or it introduces an unwanted dependence on "boundary conditions" (if the cutoff is imposed by such conditions as e.g. in [53]).

On the other hand, a strength of our approach—apart from mathematical exactness—is that postulates (I) and (II) are demonstrably satisfied for our entanglement measure E_R, but a weakness is that the basic definitions are rather indirect, and do not lead to very explicit formulas that are amenable to straightforward computations or at least approximations. In fact, as is clear from (1)–(7), we have only been able to compute upper and lower bounds. In the future, it would be interesting to establish a relation between the approaches. We conjecture that the "Buchholz free energy" $\ln Z_B(|0\rangle\langle 0|)$ introduced in the text may be seen as a regulated version of the formal quantity $\frac{1}{1-n}(\ln Z(\mathcal{M}_n) - n \ln Z(\mathcal{M}_1))$, but this remains to be investigated further. The relation between our approach and "holographic methods" based on the Ryu-Takayanagi proposal ([54, 55] and [56] for a recent textbook) is on the other hand less clear to us. Perhaps such a relation can be established via the intriguing relationship between entanglement and the geometry of the space of causal diamonds recently found by [57, 58].

The organization of this volume is as follows. For the benefit of readers with a background in Quantum Information Theory, we first review in Chap. 2 mathematical definitions and results from operator algebra theory, and present important examples of algebraic QFTs. For the benefit of readers with a background in QFT, we then review in Chap. 3 basic notions about LOCCs, entanglement measures etc., showing e.g. that (I) and (II) hold for E_R in the QFT setting. In fact, besides E_R we introduce several other such measures that serve as tools in deriving our results, and which may be of some independent interest. Our results (1)–(7) are presented in detail and proven in Chaps. 4 and 5.

Notations and conventions: Upper case Gothic letters $\mathfrak{A}, \mathfrak{B}, \ldots$ denote v. Neumann or C^*-algebras. Lower case Greek letters such as ω, σ denote linear functionals or states on a v. Neumann or C^*-algebras. Hilbert spaces are always assumed (or manifestly) separable. The dimension of spacetime is d+1, and our convention for the signature of the spacetime metric is $(- + + \ldots +)$. Scalar products on Hilbert space are anti-linear in the first entry. If $f(t), g(t)$ are non-negative functions of a real variable t, we write $f(t) \lesssim g(t)$ as $t \to \infty$ if for any $\delta > 0$ there is a t_0 such that $f(t) \leq (1 + \delta)g(t)$ for all $t \geq t_0$. We write $f(t) \sim g(t)$ when $f(t) \lesssim g(t)$ and $g(t) \lesssim f(t)$.

References

1. M. Bell, K. Gottfried, M. Veltman, *John S. Bell on the Foundations of Quantum Mechanics* (World Scientific Publishing, Singapore, 2001)
2. J.S. Bell, On the Einstein Podolsky Rosen paradox. Physics **1**, 195 (1964)
3. M.B. Plenio, S. Virmani, An introduction to entanglement measures. Quant. Inf. Comput. **7**, 1 (2007)

4. M.J. Donald, M. Horodecki, O. Rudolph, The uniqueness theorem for entanglement measures. J. Math. Phys. **43**, 4252 (2002)
5. V. Vedral, M.B. Plenio, Entanglement measures and purification procedures. Phys. Rev. A. **57**, 3 (1998)
6. H.F. Chau, C.-H. Fred Fung, H.-K. Lo, *No Superluminal Signaling Implies Unconditionally Secure Bit Commitment*, arXiv:1405.0198
7. J. Kaniewski, M. Tomamichel, E. Hänggi, S. Wehner, Secure bit commitment from relativistic constraints. IEEE Trans. Inf. Theory **59**, 4687–4699 (2013)
8. A. Kent, Quantum bit string commitment. Phys. Rev. Lett. **90**, 237901 (2003)
9. R. Haag, D. Kastler, An algebraic approach to quantum field theory. J. Math. Phys. **5**, 848 (1964)
10. R. Haag, *Local Quantum Physics: Fields, Particles, Algebras* (Springer, Berlin, 1992)
11. F.J. Murray, J. von Neumann, On rings of operators. Ann. Math. **37**(1), 116–229 (1936)
12. A. Connes, Classification of injective factors. Ann. Math. Second Ser. **104**(1), 73–115 (1976)
13. D. Buchholz, K. Fredenhagen, C. D'Antoni, The universal structure of local algebras. Commun. Math. Phys. **111**, 123 (1987)
14. K. Fredenhagen, On the modular structure of local algebras of observables. Commun. Math. Phys. **97**, 79–89 (1985)
15. R.V. Kadison, J.R. Ringrose, *Fundamentals of the Theory of Operator Algebras* (Academic Press, New York, I 1983, II 1986)
16. O. Bratteli, D.W. Robinson, *Operator Algebras and Quantum Statistical Mechanics* (Springer, I 1987, II 1997)
17. S. Doplicher, R. Longo, Standard and split inclusions of von Neumann algebras. Invent. Math. **75**, 493 (1984)
18. D. Buchholz, E.H. Wichmann, Causal independence and the energy level density of states in local quantum field theory. Commun. Math. Phys. **106**, 321 (1986)
19. J. Yngvason, Localization and engtanglement in relativistic quantum physics, in *The Message of Quantum Science*, eds. by Ph. Blanchard, J. Fröhlich. Lecture Notes in Physics, vol. 899 (Springer, Berlin, 2015), pp. 325–348
20. E. Witten, notes on some entanglement properties of quantum field theory, arXiv:1803.04993 [hep-th]
21. M. Florig, S.J. Summers, On the statistical independence of algebras of observables. J. Math. Phys. **38**, 1318 (1997)
22. H. Umegaki, Conditional expectations in an operator algebra IV (entropy and information). Kodai Math. Sem. Rep. **14**, 59–85 (1962)
23. H. Araki, Relative entropy for states of von Neumann algebras. Publ. RIMS Kyoto Univ. **11**, 809–833 (1976)
24. H. Araki, Relative entropy for states of von Neumann algebras II. Publ. RIMS Kyoto Univ. **13**, 173–192 (1977)
25. J.C. Baez, T. Fritz, A Bayesian characterization of relative entropy. Theory Appl. Categ. **29**, 421–456 (2014)
26. H. Narnhofer, Entanglement, split, and nuclearity in quantum field theory. Rep. Math. Phys. **50**, 111 (2002)
27. P. Calabrese, J. Cardy, E. Tonni, Entanglement entropy of two disjoint intervals in conformal field theory II. J. Stat. Mech. **1101**, P01021 (2011)
28. P. Calabrese, J. Cardy, E. Tonni, Entanglement entropy of two disjoint intervals in conformal field theory. J. Stat. Mech. **0911**, P11001 (2009)
29. K. Fredenhagen, K.H. Rehren, B. Schroer, Superselection sectors with braid group statistics and exchange algebras. Commun. Math. Phys. **125**, 201–226 (1989)
30. R. Longo, Index of subfactors and statistics of quantum fields. I. Commun. Math. Phys. **126**, 217 (1989)
31. R. Longo, Index of subfactors and statistics of quantum fields. 2: correspondences, braid group statistics and Jones polynomial. Commun. Math. Phys. **130**, 285 (1990)

32. R.M. Wald, *Quantum Field Theory in Curved Space-Time and Black Hole Thermodynamics* (Chicago University Press, Chicago, 1994)
33. D. Buchholz, C. D'Antoni, R. Longo, Nuclear maps and modular structures. 1. General properties. J. Funct. Anal. **88**, 223 (1990)
34. C. D'Antoni, S. Hollands, Nuclearity, local quasiequivalence and split property for Dirac quantum fields in curved space-time. Commun. Math. Phys. **261**, 133 (2006)
35. J.J. Bisognano, E.H. Wichmann, On the duality condition for quantum fields. J. Math. Phys. **17**, 303 (1976)
36. P.D. Hislop, R. Longo, Modular structure of the local algebras associated with the free massless scalar field theory. Commun. Math. Phys. **84**, 71 (1982)
37. D. Buchholz, C. D'Antoni, R. Longo, Nuclearity and thermal states in conformal field theory. Commun. Math. Phys. **270**, 267–293 (2007)
38. G. Lechner, K. Sanders, Modular nuclearity: a generally covariant perspective. Axioms **5**, 5 (2016)
39. Y. Otani, Y. Tanimoto, Towards entanglement entropy with UV cutoff in conformal nets. Ann. Henri Poincaré **19**(6), 1817–1842 (2018)
40. S. Doplicher, R. Haag, J.E. Roberts, Local observables and particle statistics. 1. Commun. Math. Phys. **23**, 199 (1971)
41. S. Doplicher, R. Haag, J.E. Roberts, Local observables and particle statistics. 2. Commun. Math. Phys. **35**, 49 (1974)
42. S. Doplicher, J.E. Roberts, Why there is a field algebra with a compact gauge group describing the superselection structure in particle physics. Commun. Math. Phys. **131**, 51 (1990)
43. S.N. Solodukhin, Entanglement entropy of black holes. Living Rev. Rel. **14**, 8 (2011)
44. T. Nishioka, S. Ryu, T. Takayanagi, Holographic entanglement entropy: an overview. J. Phys. A **42**, 504008 (2009)
45. P. Calabrese, J.L. Cardy, Entanglement entropy and quantum field theory. J. Stat. Mech. **0406**, P06002 (2004)
46. P. Calabrese, J. Cardy, Entanglement entropy and conformal field theory. J. Phys. A **42**, 504005 (2009)
47. L. Bombelli, R.K. Koul, J. Lee, R.D. Sorkin, A quantum source of entropy for black holes. Phys. Rev. D **34**, 373–383 (1986)
48. M. Srednicki, Entropy and area. Phys. Rev. Lett. **71**, 666–669 (1993)
49. L. Susskind, Some speculations about black hole entropy in string theory (1993), arXiv:hep-th/9309145 [hep-th]
50. D. Marolf, A.C. Wall, State-dependent divergences in the entanglement entropy. J. High Energy Phys. **1610**, 109 (2016)
51. H. Casini, M. Huerta, A finite entanglement entropy and the c-theorem. Phys. Lett. B **600**, 142–150 (2004)
52. H. Casini, M. Huerta, A c-theorem for the entanglement entropy. J. Phys. A **40**, 7031–7036 (2007)
53. J. Cardy, E. Tonni, Entanglement hamiltonians in two-dimensional conformal field theory. J. Stat. Mech. **1612**, 123103 (2016)
54. M. Headrick, V.E. Hubeny, A. Lawrence, M. Rangamani, Causality and holographic entanglement entropy. J. High Energy Phys. **1412**, 162 (2014)
55. S. Ryu, T. Takayanagi, Aspects of holographic entanglement entropy. J. High Energy Phys. **0608**, 045 (2006)
56. M. Rangamani, T. Takayanagi, *Holographic Entanglement Entropy*. Springer Lecture Notes in Physics (2017)
57. J. de Boer, M.P. Heller, R.C. Myers, Y. Neiman, Holographic de sitter geometry from entanglement in conformal field theory. Phys. Rev. Lett. **116**, 061602 (2016)
58. J. de Boer, F.M. Haehl, M.P. Heller, R.C. Myers, Entanglement, holography and causal diamonds. J. High Energy Phys. **1608**, 162 (2016)

Chapter 2
Formalism for QFT

Abstract This chapter presents the most important notions and examples of the theory of operator algebras. These are then used to formulate the basic principles of quantum field theory and some examples of algebraic QFTs.

2.1 C^*-Algebras and v. Neumann Algebras

This section is intended to present the most important notions from the theory of operator algebras used in later sections. (See e.g. [1, 2] for general references.) We begin with C^*-algebras:

Definition 1 A C^*-algebra is a complex, associative algebra $\mathfrak{A}$ with a unit 1, an involution $a \mapsto a^*$ and a norm $\|a\|$, such that for all $a, b \in \mathfrak{A}$, one has

$$\|ab\| \leq \|a\| \, \|b\|, \quad \|a^*\| = \|a\|, \quad \|a^*a\| = \|a\|^2, \tag{2.1}$$

and such that $\mathfrak{A}$ is complete with respect to this norm.

The norm of a C^*-algebra is an intrinsic property of the algebra in the sense that there cannot be two different C^*-norms. This is a consequence of the fact that homomorphisms between C^*-algebras, i.e. linear maps ϕ satisfying $\phi(ab) = \phi(a)\phi(b)$, $\phi(a^*) = \phi(a)^*$, are automatically continuous with $\|\phi(a)\| \leq \|a\|$. The norm of a linear functional $\varphi : \mathfrak{A} \to \mathbb{C}$ is defined by

$$\|\varphi\| = \sup_{a \in \mathfrak{A}, \|a\| \leq 1} |\varphi(a)|. \tag{2.2}$$

A linear functional on a C^*-algebra is called hermitian if $\varphi(a^*) = \overline{\varphi(a)}$ and positive if $\varphi(a^*a) \geq 0$ for all $a \in \mathfrak{A}$. A positive functional is automatically hermitian and bounded, i.e. $\|\varphi\| < \infty$. In fact, it has $\|\varphi\| = \varphi(1)$.

© The Author(s), under exclusive licence to Springer Nature Switzerland AG 2018   15
S. Hollands and K. Sanders, *Entanglement Measures and Their Properties in Quantum Field Theory*, SpringerBriefs in Mathematical Physics 34,
https://doi.org/10.1007/978-3-319-94902-4_2

Definition 2 A positive functional ω such that $\omega(1) = 1$ is called a "state".

A state automatically has $\|\omega\| = 1$, and vice-versa, any linear functional such that $\|\omega\| = 1$, $\omega(1) = 1$ is a state. A state is called "pure" if it cannot be written as a non-trivial combination $\omega = \sum_i p_i \omega_i$ of other states, where $p_i > 0$. Otherwise it is called "mixed".

A standard example of a C^*-algebra is the set $\mathfrak{B}(\mathcal{H})$ of all bounded operators on a Hilbert space $\mathcal{H}$. The norm is defined concretely in this case by the usual operator norm,

$$\|a\| = \sup_{0 \neq |\Psi\rangle \in \mathcal{H}} \| a\Psi \| / \|\Psi\| \,, \tag{2.3}$$

where the norm of a vector in Hilbert space is denoted by $\|\Psi\|^2 = \langle \Psi | \Psi \rangle$. The involution $*$ is concretely given by the hermitian adjoint, $\langle \Psi | a^* \Phi \rangle = \langle a\Psi | \Phi \rangle$ with respect to the inner product in $\mathcal{H}$. More generally, any linear subspace of $\mathfrak{B}(\mathcal{H})$ which is closed under products, hermitian adjoints, and limits, is a C^*-algebra. A *-homomorphism $\pi : \mathfrak{A} \to \mathfrak{B}(\mathcal{H})$ is called a "representation" of $\mathfrak{A}$ on $\mathcal{H}$. The statistical operators ρ on $\mathcal{H}$ (i.e. hermitian, positive semi-definite operators ρ on $\mathcal{H}$ with $\mathrm{Tr}_{\mathcal{H}}\rho = 1$) automatically give rise to states ω_ρ, in the algebraic sense of linear functionals on $\mathfrak{A}$ described above, by the formula

$$\omega_\rho(a) = \mathrm{Tr}_{\mathcal{H}}(\rho \pi(a)) \,. \tag{2.4}$$

The set of all such states is called the "folium of π", denoted $S_\pi(\mathfrak{A})$. One should be aware though that:

1. The set of states encompassed in this way by a given representation π is in general very far from containing all states ω. There can, and in general will be, disjoint folia.
2. It is in general not true that $\rho = |\Psi\rangle\langle\Psi|$ is equivalent to ω_ρ being pure!
3. It is in general not true that $\omega_\rho = \omega_{\rho'}$ implies $\rho = \rho'$.

These issues are closely related to the existence of many, often inequivalent, representations. For this, it is useful to define the notion of intertwiner. An intertwiner between two representations $(\pi_i, \mathcal{H}_i), i = 1, 2$ is a bounded linear operator $T : \mathcal{H}_1 \to \mathcal{H}_2$ such that $T\pi_1(a) = \pi_2(a)T$ for all $a \in \mathfrak{A}$. One says that

1. a representation π is irreducible if there are no non-trivial self-intertwiners (other than multiples of the identity), or equivalently, no invariant subspaces for $\pi(\mathfrak{A})$ other than $\{0\}$ and $\mathcal{H}$ itself.
2. two representations π_i are unitarily equivalent if there is a unitary intertwiner.
3. two representations π_i are quasi-equivalent if their folia coincide.
4. two representations π_i are disjoint if there is no intertwiner $T \neq 0$, i.e. if the folia $S_{\pi_i}(\mathfrak{A})$ have an empty intersection.

If the representation π is irreducible, then ω_ρ is pure if and only if $\rho = |\Psi\rangle\langle\Psi|$, and vice versa, but this is no longer true if π is not irreducible. If there are several quasi-equivalence classes, then there exist representations and states which are not represented by density matrices in this representation. Nevertheless, it can be shown that given an algebraic state $\omega : \mathfrak{A} \to \mathbb{C}$, there is always *some* representation containing a vector $|\Omega\rangle$ such that ω is represented by this vector i.e. by the density matrix $\rho = |\Omega\rangle\langle\Omega|$. This is demonstrated by a simple, canonical, but conceptually very important construction called the "GNS-construction".

The starting point of this construction is the simple observation that the algebra $\mathfrak{A}$ itself, as a linear space, always forms a representation π by left multiplication, i.e. $\pi(a)b \equiv ab$. One would like to equip this representation with a Hilbert space structure, i.e. a positive definite inner product. It seems natural to define $\langle a|b\rangle = \omega(a^*b)$, but this will in general lead to non-zero vectors with vanishing norm. Introduce $\mathfrak{J}_\omega = \{a \in \mathfrak{A} \mid \omega(a^*a) = 0\}$. By the Cauchy-Schwarz inequality, $|\omega(a^*b)| \le \omega(a^*a)^{1/2}\omega(b^*b)^{1/2}$, we have $\mathfrak{J}_\omega = \{a \in \mathfrak{A} \mid \forall b \in \mathfrak{A}, \omega(b^*a) = 0\}$, so it is a closed linear subspace and a left ideal of $\mathfrak{A}$ containing precisely the null vectors. We can then define $\mathcal{H}_\omega = \mathfrak{A}/\mathfrak{J}_\omega$ and complete it in the induced inner product. The left representation induces a representation on $\mathcal{H}_\omega$ which is called π_ω. It is the desired GNS-representation. The vector $|\Omega_\omega\rangle \in \mathcal{H}_\omega$ representing ω is simply the equivalence class of the unit operator, 1. It is by construction "cyclic" in the sense that the set $\pi_\omega(\mathfrak{A})|\Omega_\omega\rangle$ is dense in $\mathcal{H}_\omega$. We say that two states are quasi-equivalent if their GNS-representations are. Note that even mixed states are always represented by a vector in their GNS representation. Thus, in this case, the GNS-representation cannot be irreducible.

A mathematical concept related to that of a C^*-algebra is a v. Neumann algebra. Such algebras can be characterized in different ways. One way to characterize a v. Neumann algebra is as follows:

Definition 3 A v. Neumann algebra is a C^*-algebra $\mathfrak{A}$ with a distinguished folium, the folium of "normal states", which spans a linear space $S_n(\mathfrak{A})$ of linear functionals on $\mathfrak{A}$. This folium should satisfy the properties:

1. If $a, b \in \mathfrak{A}$ are such that $\omega(a) = \omega(b)$ for all $\omega \in S_n(\mathfrak{A})$, then $a = b$.
2. If $f : S_n(\mathfrak{A}) \to \mathbb{C}$ is a bounded, linear functional, then there exists an $a \in \mathfrak{A}$ such that $f(\omega) = \omega(a)$ for all ω in the distinguished folium.

One sometimes also writes $\mathfrak{A}_*$ for $S_n(\mathfrak{A})$ and calls it the "predual". States on a v. Neumann algebra which are not normal are called "singular". Given any normal state ω, we can represent $\mathfrak{A}$ on the Hilbert space $\mathcal{H}_\omega$ by the GNS-representation. The set of operators $\{\pi_\omega(a) \mid a \in \mathfrak{A}, \|a\| \le 1\}$ on $\mathcal{H}_\omega$ obtained in this way is always weakly closed, i.e. closed in the topology generated by the seminorms $N_\rho(a) = |\mathrm{Tr}(\rho\pi_\omega(a))|$.

Furthermore, the v. Neumann bi-commutant theorem holds. This theorem says the following. Let $\mathfrak{A}$ be a v. Neumann algebra represented by operators on a Hilbert space, $\mathcal{H}$, i.e. by Definition 3, $\mathfrak{A}$ can be seen as a weakly closed *-subalgebra of $\mathfrak{B}(\mathcal{H})$. The commutant is the subalgebra $\mathfrak{A}'$ of $\mathfrak{B}(\mathcal{H})$ given by all bounded operators commuting with all elements of $\mathfrak{A}$,

$$\mathfrak{A}' = \{a' \in \mathfrak{B}(\mathcal{H}) \mid [a, a'] = 0 \text{ for all } a \in \mathfrak{A}\} \ .$$

$\mathfrak{A}'$ is again a v. Neumann algebra. The v. Neumann bi-commutant theorem states that $(\mathfrak{A}')' = \mathfrak{A}'' = \mathfrak{A}$. In fact, one can show that v. Neumann algebras can actually be characterized in this way. A v. Neumann algebra represented by operators on a Hilbert space $\mathcal{H}$ is said to be in "standard form" if there exists a vector $|\Omega\rangle$ which is cyclic (see above) and separating, where "separating" means that $a|\Omega\rangle = 0$ implies $a = 0$. In our applications to quantum field theory below, the v. Neumann algebras are almost always naturally presented in such a standard form.[1]

We now come to an important construction for v. Neumann algebras in standard form due to Tomita and Takesaki. Let $S : \mathcal{H} \to \mathcal{H}$ be the anti-linear operator defined by $Sa|\Omega\rangle = a^*|\Omega\rangle$. It can be shown that the closure of S has a polar decomposition $S = J\Delta^{1/2}$, where J is anti-linear, (anti-)unitary and $\Delta^{1/2}$ is a self-adjoint, positive operator on $\mathcal{H}$. Furthermore, it can be shown that:

Proposition 1 *For any v. Neumann algebra in standard form, the operators J, $\Delta^{1/2}$ satisfy:*

1. $J\Delta^{\frac{1}{2}}J = \Delta^{-\frac{1}{2}}, \ J^* = J = J^{-1}$.
2. $\Delta|\Omega\rangle = |\Omega\rangle, \ J|\Omega\rangle = |\Omega\rangle$.
3. $a \mapsto \sigma_t(a) = \Delta^{it} a \Delta^{-it}$ *is a 1-parameter group of automorphisms of $\mathfrak{A}$.*
4. $a \mapsto JaJ$ *maps $\mathfrak{A}$ onto $\mathfrak{A}'$.*
5. *Let $\omega(a) = \langle\Omega|a\Omega\rangle$. Then ω is a **KMS-state** with respect to σ_t meaning the following. For each a, $b \in \mathfrak{A}$, there exists a complex valued function $f_{a,b}(z)$ on the strip $\{z \in \mathbb{C} \mid 0 < \Im z < 1\}$ which is bounded on the closure of the strip, analytic in the interior, and has continuous boundary values*

$$\lim_{s \to 1^-} f_{a,b}(t + is) = \omega(\sigma_{-t}(b)a), \quad \lim_{s \to 0^+} f_{a,b}(t + is) = \omega(a\sigma_{-t}(b)) \ . \quad (2.5)$$

The operators J, Δ depend on $|\Omega\rangle$ and $\mathfrak{A}$, which in principle should be included in the notation.

Key example: The essence of the theorem is maybe easiest to understand in the case when $\mathfrak{A} = M_N$ is the v. Neumann algebra of complex N by N matrices acting on $\mathbb{C}^N$. Let ω be a state on M_N. Any such state can be represented by a unique density matrix, ρ_ω, i.e. $\omega(a) = \text{Tr}_{\mathbb{C}^N}(a\rho_\omega)$. Suppose that ρ_ω has no zero eigenvalues. The GNS-representation can then be described as follows: $\mathcal{H}_\omega$ is identified with the algebra M_N itself. The GNS-vector is identified with $|\Omega_\omega\rangle = \sqrt{\rho_\omega} \in M_N$. The GNS-inner product is identified with $\langle\Psi|\Phi\rangle = \text{Tr}_{\mathbb{C}^N}(\Psi^*\Phi)$. The representation acts by left-multiplication, i.e. $\pi_\omega(a)|\Psi\rangle = |a\Psi\rangle$, $\Psi \in M_N = \mathcal{H}_\omega$. Because ρ_ω has only positive eigenvalues, it immediately follows that $|\Omega_\omega\rangle$ is separating (and cyclic), hence standard. In order to describe the operators J, Δ, we identify the Hilbert space $\mathcal{H}_\omega$ with $\mathbb{C}^N \otimes \bar{\mathbb{C}}^N$, where $\bar{\mathbb{C}}^N$ is the dual space of $\mathbb{C}^N$. Under this identification, we

---

[1] In general, a v. Neumann algebra is isomorphic to a v. Neumann algebra in standard form if it has a faithful representation which in turn is the case if it has a faithful normal state, i.e. a normal state such that $\omega(a^*a) = 0$ implies $a = 0$. In the following, we will always assume that this is the case.

may also write $\pi_\omega(a)|\Psi\rangle = (a \otimes 1)|\Psi\rangle$. The commutant, M'_N of M_N in the representation π_ω is isomorphic to the opposite algebra of M_N itself (with the products in reversed order). An element b in the commutant M'_N acts by[2] $(1 \otimes b)|\Psi\rangle$ on $\mathcal{H}_\omega$. The operators J, Δ are given by

$$\Delta^{\frac{1}{2}}|\Psi\rangle = (\rho_\omega^{1/2} \otimes \rho_\omega^{-1/2})|\Psi\rangle , \quad J|\Psi\rangle = |\Psi^*\rangle . \tag{2.6}$$

The properties (1)–(4) of the propostion are rather obvious in this example. To check (5), it is instructive to define $K = -\ln\rho_\omega$. The state may then be written as $\omega(a) = \mathrm{Tr}(ae^{-K})$, and the automorphism σ_t as $\sigma_t(a) = e^{-itK}ae^{itK}$, i.e. it corresponds to the "Heisenberg time evolution", generated by the "modular hamiltonian" K. The state ω is then obviously a Gibbs state with respect to the modular hamiltonian. The notion of KMS-state in item (5) of the previous theorem encodes precisely this. Indeed,

$$f_{a,b}(t + is) = \mathrm{Tr}(e^{-K} a e^{i(t+is)K} b e^{-i(t+is)K}) \tag{2.7}$$

has all the required properties.

With any v. Neumann algebra $\mathfrak{A}$ with standard vector $|\Omega\rangle$ there is associated a natural cone $\mathcal{P}^\sharp \subset \mathcal{H}$. It is defined by

$$\mathcal{P}^\sharp = \overline{\Delta^{\frac{1}{4}}\mathfrak{A}^+|\Omega\rangle} , \tag{2.8}$$

where $\mathfrak{A}^+$ denotes the set of non-negative elements of the v. Neumann algebra $\mathfrak{A}$, and the overbar symbol means the closure. This cone has many beautiful properties. We will need (cf. [2]):

Proposition 2   *1. Any normal state ω' has precisely one vector representative in the natural cone, i.e. $\omega'(a) = \langle \Omega'|a\Omega'\rangle$ for a unique vector $|\Omega'\rangle \in \mathcal{P}^\sharp$.*
2. *$\mathcal{P}^\sharp$ is invariant under the modular group Δ^{it}.*
3. *$\mathcal{P}^\sharp$ is the closed cone in $\mathcal{H}$ generated by vectors of the form $a(JaJ)|\Omega\rangle, a \in \mathfrak{A}$.*
4. *Let $|\Phi\rangle, |\Phi'\rangle$ be the unique vector representatives in $\mathcal{P}^\sharp$ of normal states φ, φ' on $\mathfrak{A}$. Then $\|\varphi - \varphi'\| \geq \|\Phi - \Phi'\|^2$.*

Key example continued: The meaning of this proposition is maybe best understood in the case of the previous example. According to (3), the natural cone $\mathcal{P}^\sharp$ can be seen in this example to be the set of 'vectors' in $\mathcal{H} = \mathbb{C}^N \otimes \bar{\mathbb{C}}^N$ of the form $\sum_j |\psi_j\rangle\langle\psi_j|$. A state on M_N of the form $\varphi(a) = \mathrm{Tr}_{\mathbb{C}^N}(\rho_\varphi a)$ has the vector representative $|\Phi\rangle = \sqrt{\rho_\varphi}$, which is in the cone due to the Schmidt decomposition theorem. (Alternatively, it follows from (2.6) that $\Delta^{\frac{1}{4}} = \rho_\omega^{\frac{1}{4}} \otimes \rho_\omega^{-\frac{1}{4}}$ and hence $\mathcal{P}^\sharp = \overline{\rho_\omega^{\frac{1}{4}}\mathfrak{A}^+\rho_\omega^{\frac{1}{4}}}$. Since ρ_ω is invertible, $\mathcal{P}^\sharp = \mathfrak{A}^+$, and a state φ is represented by $\sqrt{\rho_\varphi}$ in $\mathcal{P}^\sharp$.) Furthermore, in this example, the norm between two such states φ, φ' may also be written as $\|\varphi - \varphi'\| = \|\rho_{\varphi'} - \rho_\varphi\|_1$ where the norm is defined by $\|a\|_1 := \mathrm{Tr}\sqrt{a^*a}$. By the Powers-Størmer-inequality [3], we get

---

[2]It is understood here that b acts on $\langle\psi|$ in $\bar{\mathbb{C}}^N$ by $b\langle\psi| \equiv \langle b^*\psi|$.

$$\|\varphi - \varphi'\| = \|\rho_\varphi - \rho_{\varphi'}\|_1 \geq \|\sqrt{\rho_\varphi} - \sqrt{\rho_{\varphi'}}\|_2^2 , \tag{2.9}$$

where the norm $\| a \|_2 = \sqrt{\mathrm{Tr}\, a^*a}$ is the Hilbert-Schmidt norm. Under our identification $|\Phi\rangle = \sqrt{\rho_\varphi}$, $|\Phi'\rangle = \sqrt{\rho_{\varphi'}} \in \mathcal{P}^\sharp$, the Hilbert-Schmidt-norm is nothing but the Hilbert space norm in $\mathcal{H}_\omega$, so we get (4), in the case of $\mathfrak{A} = M_N$.

It is sometimes too restrictive to demand that $|\Omega\rangle$ is separating for $\mathfrak{A}$. To treat this more general situation, the above construction can be adapted as follows, see [4, 5] for details. First, one defines the subspace $\overline{\mathfrak{A}'|\Omega\rangle} = \mathcal{H}'$ with associated orthogonal projection Q onto $\mathcal{H}'$. On $Q\mathfrak{A}|\Omega\rangle \subset \mathcal{H}'$ the operator S is now defined as $SQa|\Omega\rangle = Qa^*|\Omega\rangle$ and extended by 0 on $\mathcal{H}'^\perp$, so $Sa|\Omega\rangle = Qa^*|\Omega\rangle$. The closure of S then has the decomposition $S = J\Delta^{1/2}$, and it follows that $\ker\Delta = \mathcal{H}'^\perp$ and $J^2 = Q$. As an example of this construction consider $\mathfrak{A} = M_N$, $\mathcal{H} = \mathbb{C}^N$, $|\Omega\rangle \in \mathcal{H}$. This representation is obviously irreducible, $\mathfrak{A}' = \mathbb{C}1$ and $|\Omega\rangle$ is obviously not separating. The Hilbert space $\mathcal{H}' = \mathbb{C}|\Omega\rangle$ and $\Delta = |\Omega\rangle\langle\Omega|$.

We finish this brief introduction with a subtle, but important point related to the "statistical independence" of two commuting v. Neumann algebras $\mathfrak{A}_A$ and $\mathfrak{A}_B$, represented on some common Hilbert space $\mathcal{H}$. (More precisely, we use W^*-independence in the product sense [6].) Let $\mathfrak{A}_A \vee \mathfrak{A}_B = (\mathfrak{A}'_A \cap \mathfrak{A}'_B)'$ be the v. Neumann algebra generated by $\mathfrak{A}_A$ and $\mathfrak{A}_B$ together, and let $\mathfrak{A}_A \otimes \mathfrak{A}_B$ be their v. Neumann algebraic tensor product, which we may identify with $(\mathfrak{A}_A \otimes 1) \vee (1 \otimes \mathfrak{A}_B)$ on $\mathcal{H} \otimes \mathcal{H}$.

Definition 4 The algebras $\mathfrak{A}_A$ and $\mathfrak{A}_B$ are said to be statistically independent iff there is an isomorphism of the v. Neumann algebras $\mathfrak{A}_A \vee \mathfrak{A}_B \simeq \mathfrak{A}_A \otimes \mathfrak{A}_B$.

When $\mathfrak{A}_A$ and $\mathfrak{A}_B$ are finite dimensional and $\mathfrak{A}_A \cap \mathfrak{A}_B = \mathbb{C}1$, then the algebras are always statistically independent. In the infinite dimensional case, however, statistical independence does not automatically follow. In particular, it does not follow in quantum field theory if the algebras correspond to two space-like regions which "touch each other" (see below).

Split property: The notion of statistical independence is closely related to the "split property" [7, 8]: When local algebras in quantum field theory are statistically independent, there is typically a vector $|\Psi\rangle \in \mathcal{H}$ which is cyclic for $\mathfrak{A}_A$ and $\mathfrak{A}_B$ and separating for $\mathfrak{A}_A \vee \mathfrak{A}_B$. In this case, $|\Psi\rangle \otimes |\Psi\rangle$ is cyclic and separating for $\mathfrak{A}_A \otimes \mathfrak{A}_B$ and statistical independence then entails that there is a unitary map $W : \mathcal{H} \to \mathcal{H} \otimes \mathcal{H}$ such that $(a \in \mathfrak{A}_A, b \in \mathfrak{A}_B)$

$$WaW^* = \pi_A(a) \otimes 1 , \quad WbW^* = 1 \otimes \pi_B(b) . \tag{2.10}$$

We may identify the v. Neumann algebras $\mathfrak{A}_A \simeq W\mathfrak{A}_A W^* \subset \mathfrak{B}(\mathcal{H}_A) \otimes 1$ and $\mathfrak{A}_B \simeq W\mathfrak{A}_B W^* \subset 1 \otimes \mathfrak{B}(\mathcal{H}_B)$. Furthermore, setting $\mathfrak{N} = W^*(\mathfrak{B}(\mathcal{H}_A) \otimes 1)W$, one has the inclusion

$$\mathfrak{A}_A \subset \mathfrak{N} \subset \mathfrak{A}'_B , \tag{2.11}$$

which is also called the "split". The split and the unitary W are unique (for given $|\Psi\rangle \in \mathcal{H}$) if we require that $W^*(|\Psi\rangle \otimes |\Psi\rangle)$ is in the natural cone of $|\Psi\rangle$ for $\mathfrak{A}_A \vee \mathfrak{A}_B$.

2.2 Examples of C^* and v. Neumann Algebras

We will now discuss some examples of C^* and v. Neumann algebras which are relevant in quantum physics.

2.2.1 The Weyl Algebra

The Weyl algebra encodes the canonical commutation relations. To define it we need a real vector space $K_{\mathbb{R}}$ and a symplectic form $\sigma : K_{\mathbb{R}} \times K_{\mathbb{R}} \to \mathbb{R}$. The Weyl algebra $\mathfrak{W}(K_{\mathbb{R}}, \sigma)$ is generated by elements $W(F)$, $F \in K_{\mathbb{R}}$, subject to the relations

$$W(F)W(F') = e^{\frac{-i}{2}\sigma(F,F')}W(F + F') , \quad W(F)^* = W(-F) . \tag{2.12}$$

$\mathfrak{W}(K_{\mathbb{R}}, \sigma)$ is turned into a C^*-algebra by introducing a (unique) norm and forming the completion in the norm topology, see e.g. [2, 9]. We will continue to denote this completion by $\mathfrak{W}(K_{\mathbb{R}}, \sigma)$. Due to the exponential nature of the Weyl-operators, the Weyl algebra behaves naturally under taking direct sums of symplectic vector spaces, in the sense that

$$\mathfrak{W}(K_{\mathbb{R},1} \oplus K_{\mathbb{R},2}, \sigma_1 \oplus \sigma_2) \cong \mathfrak{W}(K_{\mathbb{R},1}, \sigma_1) \otimes \mathfrak{W}(K_{\mathbb{R},2}, \sigma_2) , \tag{2.13}$$

where the precise notion of the spatial tensor product between C^*-algebras is explained in Chap. 11 of [1]. In the finite dimensional case, we may take $K_{\mathbb{R}} = \mathbb{R}^{2n}$ and σ to be the standard skew-symmetric form on $K_{\mathbb{R}}$. This gives a C^*-version of the canonical commutation relations of n positition variables and n conjugate momenta. Informally, letting $F = (p, q)$ and (P, Q) the corresponding operators with $[Q^j, P_k] = i\delta_k^j 1$, then "$W(F) = \exp[ip \cdot Q - iq \cdot P]$", and the Weyl relations formally follow by the Baker-Campbell-Hausdorff formula.

In order to obtain a v. Neumann algebra one takes a state ω and takes the weak closure in its GNS-representation, $\pi_\omega(\mathfrak{W}(K_{\mathbb{R}}, \sigma))''$. The resulting v. Neumann algebra (and even its type) will in general depend on the choice of ω. Quasifree (sometimes called "Gaussian") states of the Weyl algebra are in one-to-one correspondence with symmetric bilinear forms $\mu : K_{\mathbb{R}} \times K_{\mathbb{R}} \to \mathbb{R}$ such that

$$\mu(F, F) \geq 0 , \quad \tfrac{1}{2}|\sigma(F, F')| \leq [\mu(F, F)\mu(F', F')]^{\frac{1}{2}} \tag{2.14}$$

for all $F, F' \in K_{\mathbb{R}}$. The state corresponding to μ is defined by

$$\omega_\mu(W(F)) = e^{-\frac{1}{2}\mu(F,F)} , \tag{2.15}$$

and one checks that this is indeed positive [10]. We denote the GNS-representation by $(\pi_\mu, \mathcal{H}_\mu, \Omega)$.

The GNS-representations of quasi-free states can be described in terms of a Fock-space structure. In general, if $\mathcal{H}_1$ is a Hilbert space with inner product $(\,,\,)$, the bosonic Fock space over $\mathcal{H}_1$ is defined as the Hilbert space

$$\mathcal{F}(\mathcal{H}_1) = \mathbb{C} \oplus \bigoplus_{n>0} E_n \mathcal{H}_1^{\otimes n} \,, \tag{2.16}$$

where E_n is the projector onto the subspace of totally symmetric elements. The "vacuum" vector $|\Omega\rangle = (1, 0, 0, 0, \dots)$ corresponds to the first summand "$\mathbb{C}$". The inner product $(\,,\,)$ on this Fock space is the natural one inherited from $\mathcal{H}_1$. The summand $\mathcal{H}_1$ is called the "1-particle subspace". Creation and annihilation operators on $\mathcal{F}(\mathcal{H}_1)$ are denoted $a^*(\chi)$, $a(\chi)$, $|\chi\rangle \in \mathcal{H}_1$, respectively, where, for any $|\Psi_n\rangle = E_n |\psi_1\rangle \otimes \dots \otimes |\psi_n\rangle \in E_n \mathcal{H}_1^{\otimes n}$

$$a^*(\chi)|\Psi_n\rangle = (n+1)^{\frac{1}{2}} E_{n+1} |\chi \otimes \Psi_n\rangle \tag{2.17}$$

$$a(\chi)|\Psi_n\rangle = n^{-\frac{1}{2}} \sum_{j=1}^{n} (\chi, \psi_j) E_{n-1} |\psi_1\rangle \otimes \dots |\widehat{\psi_j}\rangle \dots \otimes |\psi_n\rangle.$$

These operators are closed and satisfy $a(\chi) = [a^*(\chi)]^*$ and $[a(\chi), a^*(\chi')] = (\chi, \chi')\mathbb{1}$.

In order to describe the GNS-representations of quasi-free states as a Fock space, it is convenient to introduce the complexification K of $K_\mathbb{R}$ and to extend μ, σ to sesquilinear forms on K. Since σ is non-degenerate, the inequality (2.14) implies that μ defines an inner product on K, which we write as $\langle\,|\,\rangle$. We denote the Hilbert space completion by $\mathrm{clo}_\mu K$. The bound (2.14) in combination with Riesz' theorem shows that there exists a unique, bounded, self-adjoint operator Σ on $\mathrm{clo}_\mu K$ such that

$$\frac{i}{2}\sigma(F, F') = \langle F, \Sigma F'\rangle \quad \text{for all } F, F' \in K. \tag{2.18}$$

We have $\|\Sigma\| \leq 1$, $\Sigma^* = \Sigma$ and $\Gamma \Sigma \Gamma = -\Sigma$, where Γ is the complex conjugation on K, which can be extended to an anti-unitary operator on $\mathrm{clo}_\mu K$. We will write the polar decomposition of Σ as $\Sigma = V|\Sigma|$, where we note that $V|\Sigma| = |\Sigma|V$ and the partial isometry V satisfies $V^* = V$.

After these preliminaries, we can now describe the GNS-representations of quasi-free states. We set $\mathcal{H}_1 := \ker(1 + \Sigma)^\perp \subset \mathrm{clo}_\mu K$. The GNS-Hilbert space of ω_μ is $\mathcal{H}_\mu = \mathcal{F}(\mathcal{H}_1)$ and the corresponding representation of the Weyl-operators $(F \in K_\mathbb{R})$ is given by

$$\pi_\mu(W(F)) = \exp[i\{a^*(\sqrt{1+\Sigma}F) + a(\sqrt{1+\Sigma}F)\}]. \tag{2.19}$$

(We refer to [2] for functional analytic details on how the exponential makes sense.) The GNS-vacuum vector is $|\Omega_\omega\rangle = (1, 0, 0, \dots)$, i.e. the Fock-space vacuum. One often informally defines the "field operator" $\phi(F) = -i\partial_t \pi_\mu(W(tF))|_{t=0}$ for $F \in K_{\mathbb{R}}$. Informally, the representation of the field operator is then

$$\phi(F) = a^*(\sqrt{1 + \Sigma} F) + a(\sqrt{1 + \Sigma}(\Gamma F)) . \tag{2.20}$$

Here we inserted the operator Γ to ensure that this field has a natural complex linear extension to $F \in K$.

Quasifree states can enjoy the following additional properties:

1. ω_μ is pure if and only if $|\Sigma| = 1$, see e.g. [11, 12], i.e. its GNS-representation is irreducible. When ω_μ is pure, then $P_\pm = \frac{1}{2}(1 \pm \Sigma)$ define projections in $\text{clo}_\mu K$ onto so-called positive and negative frequency subspaces. We then have $\mathcal{H}_1 = P_+ \text{clo}_\mu K$. Because $\Gamma P_\pm = P_\mp \Gamma$ we find $\phi(F) = \sqrt{2}\{a^*(P_+ F) + a(\Gamma P_- F)\}$.

2. ω_μ is called primary iff $\ker(\Sigma) = \{0\}$. Note that this may fail in general, because we take a completion of K. Pure states are necessarily primary. More generally, ω_μ is primary if and only if the v. Neumann algebra $\pi_\mu[\mathfrak{W}(K_{\mathbb{R}}, \sigma)]''$ of the GNS-representation has a trivial centre (i.e. it is a v. Neumann factor). In general, V^2 is the orthogonal projection onto the orthogonal complement $\ker(\Sigma)^\perp$, so $V = V^* = V^{-1}$ for primary states.

3. The GNS-vector representative $|\Omega\rangle$ is separating for $\pi_\mu[\mathfrak{W}(K_{\mathbb{R}}, \sigma)]''$ if and only if $|\Sigma| < 1$, i.e. $\ker(1 - |\Sigma|) = \{0\}$ ([13], Theorem 3.12, [14] Theorem I.3.2). In this case we have $\mathcal{H}_1 = \text{clo}_\mu K$. Note that the 1-particle space $\mathcal{H}_1$ is "twice as large" by comparison with the case of a pure state. This corresponds to the fact that the representation of $\mathfrak{W}(K_{\mathbb{R}}, \sigma)$ is now reducible.

2.2.2 The CAR Algebra

The CAR algebra encodes the canonical anti-commutation relations. To define it we need a Hilbert space K with inner product denoted by $(. , .)_K$, and an anti-linear involution Γ on K satisfying $(\Gamma k_1, \Gamma k_2)_K = (k_2, k_1)_K$. The CAR algebra $\mathfrak{C}(K, \Gamma)$ is generated by the elements $\psi(k)$, $k \in K$, subject to the relations

$$\psi(k_1)\psi(k_2) + \psi(k_2)\psi(k_1) = (\Gamma k_1, k_2)_K 1 , \quad \psi(k)^* = \psi(\Gamma k) . \tag{2.21}$$

$\mathfrak{C}(K, \Gamma)$ is turned into a C^*-algebra by introducing the (unique) norm $\|\psi(k)\| = \|k\|_K / \sqrt{2}$ and forming the completion in the norm topology, see e.g. [2, 15]. We will continue to denote this completion by $\mathfrak{C}(K, \Gamma)$. There is a *-automorphism α on $\mathfrak{C}(K, \Gamma)$ characterized uniquely by $\alpha(\psi(k)) = -\psi(k)$ which gives the CAR algebra a $\mathbb{Z}_2$-grading.

One has the functorial property

$$\mathfrak{C}(K_1 \oplus K_2, \Gamma_1 \oplus \Gamma_2) \cong \mathfrak{C}(K_1, \Gamma_1) \hat{\otimes} \mathfrak{C}(K_2, \Gamma_2) , \tag{2.22}$$

where $\hat{\otimes}$ is the graded tensor product.[3] As for the case of the Weyl algebra, one can develop a complete theory of quasi-free states and describe their representations, see [15] for details.

Here, we only describe pure, quasi-free states, and we do so by directly describing the associated GNS-representation. The input is an orthogonal projector, P, on K, obeying the relation $\Gamma P \Gamma = 1 - P$. Then one sets $\mathcal{H}_1 = PK$ equipped with the restriction $(\, . \, , \, . \,)$ of the inner product on K. Next, the fermionic Fock space over $\mathcal{H}_1$ is defined exactly as in (2.16), with the only difference that E_n now projects onto the subspace of totally anti-symmetric elements. Fermionic creation and annihilation operators are defined again as in (2.17). We then define the desired representation associated with P as

$$\pi_P(\psi(k)) = a^*(Pk) + a(P\Gamma k) . \tag{2.23}$$

The vector $|0\rangle$ in Fock space defines a state on the CAR algebra called ω_P. The "2-point" function of the state is $\omega_P(\psi(k_1)\psi(k_2)) \equiv \langle 0|\pi_P(\psi(k_1))\pi_P(\psi(k_2))|0\rangle = (\Gamma k_1, Pk_2)$, and similar formulas can be derived for the higher "n-point" functions, see [15] for details.

2.2.3   The Cuntz Algebra $\mathcal{O}_n$

A C^*-algebra arising naturally in the theory of superselection sectors (see below Sect. 4.7) is the Cuntz-algebra. Let $\mathcal{H}$ be a separable, infinite dimensional Hilbert space. A partial isometry, V, on $\mathcal{H}$ is a linear operator such that $V^*V = 1$ and such that VV^* is a projection. Now let $n > 1$ a natural number. Then it is not hard to see that one can construct partial isometries $\psi_i, i = 1, \ldots, n$ on $\mathcal{H}$ satisfying the relations

$$\sum_{i=1}^{n} \psi_i \psi_i^* = 1 , \quad \psi_i^* \psi_j = \delta_{ij} 1 \quad \text{for all } i, j. \tag{2.24}$$

Cuntz [16] has shown that there exists a unique (up to C^*-isomorphism) simple C^*-algebra generated by these elements and relations. It is denoted $\mathcal{O}_n$.

---

[3] As a vector space, the graded tensor product is first defined to be the usual (algebraic) tensor product. The product is defined as $(a_1 \hat{\otimes} b_1)(a_2 \hat{\otimes} b_2) = (-1)^{deg(a_2)deg(b_1)} a_1 a_2 \hat{\otimes} b_1 b_2$ and the *-operation is $(a \hat{\otimes} b)^* = (-1)^{deg(a)deg(b)} a^* \hat{\otimes} b^*$, where the degree is defined to be 0 resp. 1 for even resp. odd elements under α. It is then shown that a natural C^*-norm compatible with these relations and the above isomorphism can be defined which extends the C^*-norm of $\mathfrak{C}(K_i, \Gamma_i)$. The graded tensor product is the C^*-closure under this norm.

2.3 The Basic Principles of Quantum Field Theory

In algebraic quantum field theory, the algebraic relations between the quantum fields are encoded in a collection of C^* or v. Neumann algebras associated with spacetime regions. The precise framework depends somewhat on the type of theory, spacetime background etc. one would like to consider. In the case of Minkowski space $\mathbb{R}^{d,1}$, a standard set of assumptions, manifestly satisfied by many examples, and believed to be satisfied by all reasonable QFTs, is as follows. Call a "causal diamond" $O \subset \mathbb{R}^{d,1}$ any set of the form $O = D(A)$, where A is any relatively compact open set contained in a Cauchy surface $\cong \mathbb{R}^d$ of Minkowski space, and $D(A)$ its domain of dependence, i.e. the set of points $x \in \mathbb{R}^{d,1}$ such that any inextendible causal curve through x must hit A at least once, see Fig. 1.1. Poincaré transformations $g = (\Lambda, a) \in \mathrm{P} = \mathrm{SO}_+(d, 1) \ltimes \mathbb{R}^{d+1}$ act on points by $g \cdot x = \Lambda x + a$. Since Poincaré transformations are isometries of Minkowski spacetime, they map causal diamonds to causal diamonds, so we get an action $O \mapsto g \cdot O$ on the set of causal diamonds.

In the algebraic approach, a quantum field theory is a collection ("net") of C^*-algebras $\mathfrak{A}(O)$ subject to the following conditions:

(a1) (Isotony) $\mathfrak{A}(O_1) \subset \mathfrak{A}(O_2)$ if $O_1 \subset O_2$. We write $\mathfrak{A} = \overline{\bigcup_O \mathfrak{A}(O)}$ with completion in the C^*-norm.

(a2) (Causality) $[\mathfrak{A}(O_1), \mathfrak{A}(O_2)] = \{0\}$ if O_1 is space-like related to O_2. In other words, algebras for space-like related causal diamonds commute. Denoting the causal complement of a set O by O', we may also write this more suggestively as

$$\mathfrak{A}(O') \subset \mathfrak{A}(O)'$$

where the prime on the right side is the commutant.

(a3) (Relativistic covariance) For each transformation $g \in \widetilde{\mathrm{P}}$ covering[4] a Poincaré transformation $(\Lambda, a) \in \mathrm{P} = \mathrm{SO}_+(d, 1) \ltimes \mathbb{R}^{d+1}$, there is an automorphism α_g on $\mathfrak{A}$ such that $\alpha_g \mathfrak{A}(O) = \mathfrak{A}(\Lambda O + a)$ for all causal diamonds O and such that $\alpha_g \alpha_{g'} = \alpha_{gg'}$ and $\alpha_{(1,0)} = \mathrm{id}$ is the identity.

(a4) (Vacuum) There is a unique state ω_0 on $\mathfrak{A}$ invariant under α_g. On its GNS-representation $(\pi_0, \mathcal{H}_0, |0\rangle)$, α_g is implemented by a projective positive energy representation U of the Poincaré group $\widetilde{\mathrm{P}}$ in the sense that $\pi_0(\alpha_g(a)) = U(g)\pi_0(a)U(g)^*$ for all $a \in \mathfrak{A}$, $g \in \widetilde{\mathrm{P}}$. Positive energy means that the representation is strongly continuous and that, if $x \in \mathbb{R}^{d,1} \subset \mathrm{P}$ is a translation by x, we can write

$$U(x) = \exp(-iP^\mu x_\mu), \tag{2.25}$$

and the vector generator $P = (P^\mu)$ has spectral values $p = (p^\mu)$ in the forward lightcone $p \in \bar{V}^+ = \{p \mid p^2 \geq 0, p^0 > 0\}$.

---

[4] The covering group is needed to describe non-integer spin.

For technical reasons, one often forms the weak closures $[\pi_0(\mathfrak{A}(O))]''$, which gives a corresponding net of v. Neumann algebras (on $\mathcal{H}_0$). We will often work with this net. Axioms (a1) and (a2) can be generalized straightforwardly to any general globally hyperbolic spacetime ($\mathcal{M}$, g). If such a spacetime has any symmetries, then (a3) can be generalized too, with the Poincaré group replaced by the isometry group G of the spacetime. If this group G has a 1-parameter family of isometries with everywhere complete time-like orbits (as is the case e.g. in a static spacetime), then a version of (a4) can be imposed, too. For more details, we refer the reader to [17], and references therein.

A straightforward, but important, consequence of axioms (a1)–(a4) (and Araki's "weak additivity axiom" [18], see below) is the Reeh-Schlieder theorem [19], which is the following. We know by construction that $\pi_0(\mathfrak{A})|0\rangle$ is dense in the entire Hilbert space, $\mathcal{H}_0$. One might guess at first that the subspace of states $\overline{\pi_0(\mathfrak{A}(O))}|0\rangle$, describing excitations relative to the vacuum localized in a causal diamond O, would depend on O. This expectation is usually incorrect, however, and instead the Reeh-Schlieder theorem holds:

Theorem 1 (Reeh-Schlieder theorem) *Assume that any element of $\mathfrak{A}$ can be approximated arbitrarily well in the sense of matrix elements in the vacuum representation π_0 by finite sums of elements of the form $\alpha_{x_i}(a_i)$, where a_i are in some arbitrarily small causal diamond, and where α_{x_i} denotes a translation ("weak addititivity[5] "). For any causal diamond O, the set of vectors $\pi_0(\mathfrak{A}(O))|0\rangle$ is dense in the entire Hilbert space. The same statement remains true if $|0\rangle$ is replaced with a vector with finite energy.*

Proof The proof of this theorem is standard. We drop the reference to π_0, so a means $\pi_0(a)$, etc. Suppose that the closure of $\{a|0\rangle \mid a \in \pi_0(\mathfrak{A}(\tilde{O}))\}$ has a non-trivial orthogonal complement in $\mathcal{H}_0$ where $\tilde{O}$ is a causal diamond whose closure is contained in O (i.e. "inside" O). Let $|\Phi\rangle$ be a vector in the orthogonal complement, $a \in \mathfrak{A}(\tilde{O})$, and let $\mathbb{R}^{d,1} \ni z \mapsto f(z) = \langle\Phi|U(z)a|0\rangle$. Due to the spectral properties of P postulated in (a3), $f(z)$ has a holomorphic extension to the set $\{z \in \mathbb{C}^{d+1} \mid \Im(z) \in V^+\}$. By assumption $f(z) = 0$ if z is in some sufficiently small real neighborhood of $z = 0$, since the translated element $U(z)aU(z)^*$ remains in $\mathfrak{A}(O)$ for such z, by (a3). By the edge-of-the-wedge theorem, see Appendix A.1, $f(z) = \langle\Phi|U(z)aU(z)^*|0\rangle$ therefore vanishes for all $z \in \mathbb{R}^{d,1}$. Since an arbitrary $b \in \mathfrak{A}$ can be approximated in the sense of matrix elements by a sum of translated elements from $\mathfrak{A}(\tilde{O})$ (by assumption), we have $\langle\Phi|b|0\rangle = 0$ for all $b \in \mathfrak{A}$, proving that also $|\Phi\rangle$ vanishes identically.

If $|\Psi\rangle$ is a vector with finite energy, then $U(z)^*|\Psi\rangle$ has an analytic continuation to $\{z \in \mathbb{C}^{d+1} \mid \Im(z) \in V^+\}$ and a similar argument applies.   □

---

[5]In terms of algebras:

$$\pi_0(\mathfrak{A}) = \left(\bigcup_{x \in \mathbb{R}^{d,1}} \pi_0(\mathfrak{A}(O + x))\right)''$$

for any causal diamond O.

The Reeh-Schlieder theorem shows in particular that it is not straightforward to define "localized states". To define these, and to have a mathematical counterpart of the intuitive idea that the set of states localized in O with energy below E should be approximately finite-dimensional, one has to go beyond the above axioms [20]. Nowadays, this idea is phrased mathematically in terms of "nuclearity conditions", which will also play a role in this work. One such condition, due to Buchholz and Wichmann [8], is described now. We formulate it as an additional axiom (a5), although the precise form of this axiom is perhaps not so natural and irrefutable as the previous ones, (a1)–(a4), and in fact, we will throughout this volume also consider variants of (a5) (see (a5') in Chap. 4).

(a5) (BW-nuclearity) Let A be a ball of radius r in a time-slice, and let $O_r = D(A)$ be the corresponding double cone. Consider the map

$$\Theta_{\beta,r} : \mathfrak{A}(O_r) \to \mathcal{H}_0 , \quad a \mapsto e^{-\beta H} \pi_0(a)|0\rangle , \tag{2.26}$$

where $\beta > 0$ and where $H = P^0$ is the Hamiltonian, i.e. the time-component of P^μ in item (a4). It is required that there exist positive constants $n > 0$ and $c = c(r) > 0$ such that for $r > 0$, $\beta > 0$

$$\|\Theta_{\beta,r}\|_1 \le e^{(c/\beta)^n} . \tag{2.27}$$

Here we use the following [21, 22] definition/lemma:

Definition 5 The 1-nuclear norm $\| \cdot \|_1$ of a linear operator T between two Banach spaces $\mathcal{X}, \mathcal{Y}$ is defined as $\|T\|_1 = \inf \sum_j \|y_j\| \, \|\psi_j\|$, where the infimum is taken over all possible ways (if any[6]) of writing $T(\,.\,) = \sum_j y_j \psi_j(\,.\,)$ as a norm convergent sum in terms of linear functionals $\psi_j \in \mathcal{X}^*$ and vectors $y_j \in \mathcal{Y}$.

If both $\mathcal{X}, \mathcal{Y}$ are Hilbert spaces, then $\|T\|_1 = \text{Tr}|T|$. The 1-nuclear norm satisfies the following properties:

$$\|ST\|_1 \le \|S\| \|T\|_1 , \quad \|ST\|_1 \le \|S\|_1 \|T\| , \tag{2.28}$$

where S is another linear operator between Banach spaces $\mathcal{Y}, \mathcal{Z}$.

The idea is that this 1-norm measures the "size" of the ellipsoid in Hilbert space given by the collection of vectors of the form $e^{-\beta H} \pi_0(a)|0\rangle$ with $\|a\| \le 1$. Such vectors represent states which are localized in space (by r) and have an exponentially suppressed energy (essentially by $1/\beta$). One typically expects $c \propto r$ and $n = d$ in d spatial dimensions. (Note that in quantum mechanics, using $\mathcal{B}(\mathcal{H})$ instead of $\mathfrak{A}(O_r)$ and assuming that H has an orthonormal eigenbasis, $\|\Theta_{\beta,r}\|_1$ reduces to the partition function.)

The BW-nuclearity condition (a5) and its variants—in particular the modular nuclearity condition (a5') given below—can be shown to have a consequence which

---

[6]If there are none, then the norm is set to infinity.

is of essential importance in this volume. If A and B are disjoint regions in a spatial slice with distance $\text{dist}(A, B) > 0$ and corresponding causal diamonds O_A and O_B, then (a5) implies that the split property holds for $\mathfrak{A}_A = \pi_0(\mathfrak{A}(O_A))''$ and $\mathfrak{A}_B = \pi_0(\mathfrak{A}(O_B))''$ [23]. In other words, the v. Neumann algebras not only commute (by item (a2)), but they are statistically independent. This fact will be crucial when defining relative entanglement entropies between such algebras.

A closely related consequence of (a5) describes what happens when the regions O_A and O_B "touch" each other. It deals with a net of v. Neumann algebras $\mathfrak{A}(O)$, and it assumes an additional asymptotic scale invariance at small length scales. Under these circumstances, the local v. Neumann algebras are direct sums of type III_1 factors [23]. Furthermore, when we call a state ω on $\mathfrak{A}$ locally normal when its restriction to each $\mathfrak{A}(O)$ is normal, we have the following [24]:

Theorem 2 (a) *The restriction of a locally normal state ω on $\mathfrak{A}$ to a local algebra $\mathfrak{A}(O)$ is* never *pure.*

(b) *If ω_O is a pure normal state on some local algebra $\mathfrak{A}(O)$ then it cannot be extended to a normal state on any larger local algebra.*

(c) *Let A and B be disjoint subsets of a Cauchy surface touching each other in the sense that $\bar{A}$ intersects $\bar{B}$. Then there cannot exist a normal, separable state on $\mathfrak{A}(O_A) \vee \mathfrak{A}(O_B)$, i.e. one such that $\omega(ab) = \sum_j \varphi_j(a)\psi_j(b)$ for all $a \in \mathfrak{A}(O_A), b \in \mathfrak{A}(O_B)$, where φ_j, ψ_j are normal positive functionals on $\mathfrak{A}(O_A), \mathfrak{A}(O_B)$, respectively. In particular, $\mathfrak{A}(O_A), \mathfrak{A}(O_B)$ cannot be statistically independent.*

Proof (a) follows immediately from (b), which is proved in Corollary 3.3 of [24]. To see that (c) follows from (a) we assume that a normal separable state ω exists, and we consider its separable decomposition. We can then find pure normal states ω_A and ω_B on $\mathfrak{A}(O_A)$ and $\mathfrak{A}(O_B)$, respectively, and a constant $c > 0$ such that $\omega_A \leq \sqrt{c}\varphi_1$ and $\omega_B \leq \sqrt{c}\psi_1$. For all $a \in \mathfrak{A}(O_A)$ and $b \in \mathfrak{A}(O_B)$ we then have $0 \leq \omega_A(a^*a)\omega_B(b^*b) \leq c\omega(a^*ab^*b)$, an estimate which can be extended to the algebraic tensor product $\mathfrak{A}(O_A) \odot \mathfrak{A}(O_B)$ and then to the entire algebra $\mathfrak{A}(O_A) \vee \mathfrak{A}(O_B)$. We denote the resulting state constructed out of ω_A and ω_B with a tensor product, so that $\omega_A \otimes \omega_B \leq c\omega$. This estimate shows that $\omega_A \otimes \omega_B$ must be normal on $\mathfrak{A}(O_A) \vee \mathfrak{A}(O_B)$, which means in turn that it can be described by a density matrix ρ_{AB} on the vacuum Hilbert space. Finally, ρ_{AB} defines a locally normal state on all of $\mathfrak{A}$, which restricts to the pure normal state ω_A on $\mathfrak{A}(O_A)$, contradicting statement (a). Hence, a normal separable state on $\mathfrak{A}(O_A) \vee \mathfrak{A}(O_B)$ cannot exist. $\qquad\square$

The BW-nuclearity condition is designed specifically for Minkowski space, or more generally for a spacetime with a 1-parameter group of time-like isometries, where a Hamiltonian exists. There are also other nuclearity conditions allowing to draw similar conclusions, most notably ones involving the modular operator Δ [25, 26] introduced below as (a5') in Chap. 4.

2.4 Examples of Algebraic Quantum Field Theories

Let us now discuss some examples of algebraic quantum field theories which manifestly fit into the operator algebraic setting.

2.4.1 Free Scalar Fields

The free Klein-Gordon (KG) field ϕ of mass m on $d+1$-dimensional Minkowski spacetime (or, more generally, on any globally hyperbolic manifold $(\mathcal{M}, g)$) is the simplest example satisfying all the axioms, including the BW-nuclearity condition, cf. [8, 27]. Roughly speaking, $\phi(x)$ is an operator for each $x \in \mathcal{M}$ satisfying the KG-equation $(\Box - m^2)\phi(x) = 0$. If $Q = \phi|_{\mathcal{C}}$, $P = \partial_n \phi|_{\mathcal{C}}$ are the restriction of the field and its normal derivative to a Cauchy surface ("time-slice") $\mathcal{C}$, then the canonical commutation relations $[Q(x), P(y)] = i\delta_{\mathcal{C}}(x, y)\mathbb{1}$ are satisfied, and, informally, the algebra $\mathfrak{A}(O_A)$ with $A \subset \mathcal{C}$, should be thought of as being generated by all $P(x)$, $Q(x)$ with $x \in A$. For a mathematically precise construction one has to pass to bounded operators, which can be done e.g. by treating the field operators $\phi(x)$ as a distribution and by working with suitable exponentials of smeared versions of these fields.

This construction is as follows. We let $K_{\mathbb{R}} = C_0^\infty(\mathcal{C}, \mathbb{R}) \oplus C_0^\infty(\mathcal{C}, \mathbb{R})$, and we define a symplectic form, corresponding to the canonical commutation relations, by $\sigma(F, F') = \int_{\mathcal{C}}[qp' - q'p]\mathrm{d}V$ for all $F = (q, p)$, $F' = (q', p') \in K_{\mathbb{R}}$. For $A \subset \mathcal{C}$, we define $\mathfrak{A}(O_A)$ to be the subalgebra of the Weyl algebra $\mathfrak{W}(K_{\mathbb{R}}, \sigma)$ given by the C^*-closure of $\{W(F) \mid \mathrm{supp}(F) \subset A\}$.[7] We think of the Weyl operators $W(F)$ as representing the informal expressions $\exp[i \int_{\mathcal{C}}(pQ - qP)\mathrm{d}V]$.

This construction has the disadvantage that we only identify the algebras corresponding to causal diamonds O_A with base A in the fixed Cauchy-surface $\mathcal{C}$, see Fig. 1.1. However, there is a simple way to get around this by giving an alternative description of $(K_{\mathbb{R}}, \sigma)$. For this, we note that $K_{\mathbb{R}}$ is just the space of initial data of the KG equation, equipped with the natural symplectic form. Since initial data are, on a globally hyperbolic spacetime, in one-to-one correspondence with solutions to the KG equation, we may equally well work with solutions. For this, let G^A, $G^R : C_0^\infty(\mathcal{M}, \mathbb{R}) \to C^\infty(\mathcal{M}, \mathbb{R})$ be the unique advanced and retarded propagators to the KG equation uniquely specified by the relations

$$(\Box - m^2)G^{A,R} = 1_M, \quad \mathrm{supp}(G^{A,R}f) \subset J^{-,+}(\mathrm{supp}(f)), \qquad (2.29)$$

where $J^{\pm}$ denotes the causal future/past of a set in the spacetime $(\mathcal{M}, g)$, see e.g. [28]. Let $G = G^R - G^A$ be the "causal propagator". It can be identified with a distributional kernel on $\mathcal{M} \times \mathcal{M}$, denoted by $G(x, y)$. This distribution is real-

---

[7]Here the braces denote "generated by, as a C^*-algebra", and $\mathrm{supp}(F) = \mathrm{supp}(q) \cup \mathrm{supp}(p)$, where the support supp of a function is the closure of the set of all points where it does not vanish.

valued, anti-symmetric, a solution to the KG equation in both entries, and vanishes if x is space-like to y. Define $\tilde{K}_\mathbb{R} = C_0^\infty(\mathcal{M}, \mathbb{R})/\mathrm{Ran}(\square - m^2)$ (with $(\square - m^2)$ acting on $C_0^\infty(\mathcal{M}, \mathbb{R})$). Elements in this space are equivalence classes $[f]$ of compactly supported, smooth, real-valued functions f on $\mathcal{M}$ modulo ones in the image of the KG operator. On $\tilde{K}_\mathbb{R}$ we define a symplectic form by $\tilde{\sigma}([f], [f']) = G(f, f')$. One now shows that $(\tilde{K}_\mathbb{R}, \tilde{\sigma})$ is a symplectic space isomorphic to $(K_\mathbb{R}, \sigma)$. The isomorphism is given by assigning to $[f] \in \tilde{K}_\mathbb{R}$ the initial data $(q, p) \in K_\mathbb{R}$ for the solution Gf,

$$[f] \mapsto \begin{pmatrix} Gf|_C \\ \partial_n Gf|_C \end{pmatrix} \equiv \begin{pmatrix} q \\ p \end{pmatrix}. \tag{2.30}$$

Consequently, the Weyl algebra $\mathfrak{W}(\tilde{K}_\mathbb{R}, \tilde{\sigma})$ is isomorphic to $\mathfrak{W}(K_\mathbb{R}, \sigma)$. The local algebras for an arbitrary $O \subset \mathcal{M}$ are now defined as

$$\mathfrak{A}(O) = \{\tilde{W}([f]) \mid \mathrm{supp}(f) \subset O\}. \tag{2.31}$$

The obvious analogs of axioms (a1), (a2) for a Lorentzian spacetime $(\mathcal{M}, g)$ directly follow from the properties of G and the relation $\tilde{W}([f])\tilde{W}([f]) = e^{-iG(f,f')/2}\tilde{W}([f + f'])$. Given a representation π of the Weyl algebra on a Hilbert space, one may informally think of $\pi(\tilde{W}([f]))$ as $e^{i\phi(f)}$, where $\phi(f) = \int_M \phi(x)f(x)dV$ is a "smeared" quantized Klein-Gordon field. The field $\phi(f)$ actually exists as a self-adjoint, operator valued distribution in general only in certain representations of the Weyl-algebra.

One such representation is the ground state representation of a massive ($m > 0$) field on an ultra-static spacetime $\mathcal{M} = \mathbb{R} \times C$, $g = -dt^2 + h$, where h is a Riemannian metric on C not depending on t, and where C is assumed to be compact. Let ∇^2 be the Laplacian of the Riemannian metric h on C. $-\nabla^2$ is known to be a positive essentially self-adjoint operator on $L^2(C)$ [29], so in particular all complex powers of $-\nabla^2 + m^2$ are well-defined. Let $C = (-\nabla^2 + m^2)^{-1}$. Consider the bilinear form $\mu : K_\mathbb{R} \times K_\mathbb{R} \to \mathbb{R}$ given by

$$\mu(F, F') = \tfrac{1}{2}\Re(C^{1/4}p - iC^{-1/4}q, C^{1/4}p' - iC^{-1/4}q') \tag{2.32}$$

where $F = (q, p)$ and we mean the standard L^2-inner product on the right hand side. It follows immediately that (2.14) holds, so μ defines a quasi-free state ω_0 on $\mathfrak{W}(K_\mathbb{R}, \sigma)$, and therefore also one on $\mathfrak{W}(\tilde{K}_\mathbb{R}, \tilde{\sigma})$ via the isomorphism described above in (2.30). The closure of $K_\mathbb{R}$ in the topology given by μ is isomorphic to $W_\mathbb{R}^{+1/2}(C) \oplus W_\mathbb{R}^{-1/2}(C)$ (here we mean the Sobolev spaces of fractional order $\pm 1/2$) and the corresponding operator Σ as defined in Eq. (2.18) is given by

$$\Sigma = i \begin{pmatrix} 0 & C^{1/2} \\ -C^{-1/2} & 0 \end{pmatrix}. \tag{2.33}$$

It follows that $\Sigma^2 = 1$, so the state ω_0 is pure. If we compose the map $\sqrt{1 + \Sigma}$ with the identification $W^{+1/2}(\mathcal{C}) \oplus W^{-1/2}(\mathcal{C}) \simeq L^2(\mathcal{C}) \oplus L^2(\mathcal{C})$, determined by $(q, p) \mapsto \frac{1}{\sqrt{2}}(C^{-\frac{1}{4}}q, C^{\frac{1}{4}}p)$, we obtain

$$K_{\mathbb{R}} \ni F = \begin{pmatrix} q \\ p \end{pmatrix} \mapsto \frac{1}{2} \begin{pmatrix} iC^{1/4}p + C^{-1/4}q \\ C^{1/4}p - iC^{-1/4}q \end{pmatrix} \equiv \frac{1}{\sqrt{2}} \begin{pmatrix} i\kappa(F) \\ \kappa(F) \end{pmatrix} \in L^2(\mathcal{C})^{\oplus 2}. \quad (2.34)$$

From this it immediately follows that the image of $K_{\mathbb{R}}$ under $\sqrt{1 + \Sigma}$ is isomorphic to $L^2(\mathcal{C})$, which may thus be identified with the 1-particle Hilbert space $\mathcal{H}_1$. The map $\kappa : K_{\mathbb{R}} \to L^2(\mathcal{C}) = \mathcal{H}_1$ defined in the previous equation is the one-particle structure of Kay [30]. In terms of κ one may write the GNS-representation of our state ω_0 as ($F \in K_{\mathbb{R}}$)

$$\pi_0(W(F)) = \exp i[a^*(\kappa(F)) + a(\kappa(F))], \quad (2.35)$$

where a, a^* are annihilation and creation operators on the bosonic Fock space over $\mathcal{H}_1$. One can show (cf. [30]), that the time translation isometry of $(\mathcal{M}, g)$ is implemented in the GNS representation of this state by a strongly continuous 1-parameter group with positive generator H (the Hamiltonian). Thus, the state can be considered as a ground state. The same construction works in any ultra-static, globally hyperbolic spacetime even when the slices are not compact (because $\mathcal{C}$ is then complete). In particular, it works in Minkowski space, where the Cauchy surface is given by $\mathcal{C} = \mathbb{R}^d$, equipped with the flat metric.

In Minkowski space, it is instructive to represent 1-particle wave functions in $\mathcal{H}_1$ in momentum space, using the Fourier transform[8] to identify $L^2(\mathbb{R}^d, d^d\mathbf{x})$ with $L^2(\mathbb{R}^d, \frac{1}{2\omega(\mathbf{k})}d^d\mathbf{k})$, where $\omega(\mathbf{k}) := \sqrt{\mathbf{k}^2 + m^2}$. More precisely, we map $\psi(x)$ to $\sqrt{2\omega(\mathbf{k})}\tilde{\psi}(\mathbf{k})$, which defines an isometry. For any smooth function f on Minkowski space of compact support, one then writes the smeared quantum KG-field characterized by $\pi_0(\tilde{W}([f])) = \exp i\phi(f)$ informally as an integral $\int \phi(x) f(x) d^{d+1}x$. Going through the definitions and constructions just given, one can then further write the representer in the familiar form

$$\phi(x) = (2\pi)^{-\frac{d}{2}} \int_{\mathbb{R}^d} \frac{d^d\mathbf{k}}{2\omega(\mathbf{k})} [a^\dagger(\mathbf{k})e^{-ikx} + a(\mathbf{k})e^{ikx}] \quad (2.36)$$

in terms of the creation and annihilation operators[9] on the bosonic Fock space over $\mathcal{H}_1 = L^2(\mathbb{R}^d, \frac{1}{2\omega(\mathbf{k})}d^d\mathbf{k})$, where $xk = x^\mu k_\mu$ and $(k_\mu) = (-\omega(\mathbf{k}), \mathbf{k})$. After smearing with $f \in C_0^\infty(\mathbb{R}^{d+1})$, this operator is shown to have a dense set of analytic vectors, so it is essentially self-adjoint by Nelson's analytic vector theorem [31]. Thus, we

---

[8] Our convention for the Fourier transform in one dimension is $\tilde{f}(p) = \frac{1}{\sqrt{2\pi}} \int dx f(x)e^{-ipx}$.

[9] In our setup, we arrive at the normalization

$$[a(\mathbf{k}), a^\dagger(\mathbf{k}')] = 2\omega(\mathbf{k}) \delta^d(\mathbf{k} - \mathbf{k}') \cdot 1, \quad [a(\mathbf{k}), a(\mathbf{k}')] = 0. \quad (2.37)$$

may write the representation of the Weyl-operators by exponentiation in the form $\pi_0(\tilde{W}([f])) = \exp[i\phi(f)]$, so the relation between Weyl operators and unbounded quantum fields is more than a formal one in this example. The vacuum state satisfies the axioms (a3), (a4) as well as the BW-nuclearity condition [8].

2.4.2 Free Fermion Fields

The free spin-$\frac{1}{2}$-field ψ of mass m on $d+1$-dimensional Minkowski spacetime (or, more generally, on any globally hyperbolic spin manifold $(\mathcal{M}, g)$) is another simple example satisfying all the axioms, including the BW-nuclearity condition in static spacetimes [32]. Roughly speaking, $\psi(x)$ is an operator for each $x \in \mathcal{M}$ satisfying the Dirac equation $(\slashed{\nabla} + m)\psi(x) = 0$. For a mathematically precise construction one uses the CAR algebra described in Sect. 2.2.2.

This construction is as follows e.g. for a real (Majorana) field, which exists when $d - 1 = 0, 1, 2, 7 \bmod 8$, for details see e.g. [32]. Let \$ be the spin-bundle, which is a complex vector bundle over $\mathcal{M}$ of dimension $2^{\lfloor(d+1)/2\rfloor}$. At each point, we let e_μ be a chosen orthonormal $d + 1$-bein, so that $\slashed{\nabla} = e_\mu \cdot \nabla_{e_\mu}$ is the Dirac operator, with $e_\mu\cdot$ denoting Clifford multiplication. The action of e_μ on the conjugate vector bundle[10] $\bar{\$}$ is denoted by $\overline{e_\mu}\cdot$. There is then an anti-linear map $C : \$ \to \bar{\$}$ satisfying $\overline{C}C = (-1)^{\frac{1}{2}(d-1)d}$ which is characterized by $C^{-1}\overline{e}_\mu C = e_\mu$. We also have "Dirac conjugation", which is an anti-linear map $B : \$ \to \$^*$ with $B^*B = 1$ (for details on such notions, see e.g. [33]). The Hilbert space K needed for the definition of the CAR algebra $\mathfrak{C}(K, \Gamma)$ is the space $K = L^2(\mathcal{C}, \$|_\mathcal{C})$ of square integrable spinors on a Cauchy surface $\mathcal{C}$ (unit forward normal: n), with inner product

$$(k_1, k_2)_K = \int_\mathcal{C} Bk_1(n \cdot k_2)\, \mathrm{d}V \qquad (2.38)$$

and the anti-linear involution Γ is given by $\Gamma k = \overline{C}\overline{k}$. If $A \subset \mathcal{C}$ and O_A its causal diamond, we define

$$\mathfrak{A}(O_A) = \{\psi(k) \mid k \in K, \mathrm{supp}(k) \subset A\}^{\mathrm{even}}, \qquad (2.39)$$

where on the right side, curly brackets mean "generated by, as a C^*-algebra", and the superscript denotes the subalgebra of elements with an even number of generators.

As before, this construction has the disadvantage that only the algebras corresponding to causal diamonds on a fixed Cauchy-surface $\mathcal{C}$ are identified. However, there is a simple way to get around this by giving an alternative description of (K, Γ). For this, we note that K is just the space of initial data of the Dirac equation, equipped with the natural hermitian inner product. Since initial data are,

---

[10]The complex conjugate, $\bar{V}$, of a vector space V is identical as a set, but has the scalar multiplication $\lambda \cdot v \equiv \bar{\lambda}v$.

on a globally hyperbolic spacetime, in one-to-one correspondence with solutions to the Dirac equation, we may equally well work with solutions. For this, let $\rlap{E}{/}^A$, $\rlap{E}{/}^R : C_0^\infty(\mathcal{M}, \$) \to C^\infty(\mathcal{M}, \$)$ be the unique advanced and retarded propagators associated with the "squared" Dirac operator

$$(\nabla\!\!\!\!/ + m)(\nabla\!\!\!\!/ - m) = \Box - m^2 - \frac{1}{4}R . \tag{2.40}$$

The advanced propagator for the Dirac equation is simply $\$^A = (\nabla\!\!\!\!/ - m)\rlap{E}{/}^A$ (and similarly for the retarded propagator), and we also set $\$ = \$^A - \$^R$. We next equip the complex vector space of all smooth compactly supported spinors with the hermitian sesquilinear form

$$(k_1, k_2)_{\tilde{K}} = \int_{\mathcal{M}} Bk_1(\$k_2) . \tag{2.41}$$

This sesquilinear form is checked to be positive semi-definite, and after factoring with the kernel of $\$$, it becomes a pre-Hilbert space $\tilde{K} = C_0^\infty(\mathcal{M}, \$)/\mathrm{Ran}(\nabla\!\!\!\!/ + m)$. Now one shows that the closure of this Hilbert space, with the above inner product and conjugation $\tilde{\Gamma} = \Gamma$ is in fact isometric to (K, Γ) as defined before under restriction to C, i.e. under $[k] \in \tilde{K} \mapsto k|_C \in K$. Consequently, the CAR algebra $\mathfrak{C}(K, \Gamma)$ is isomorphic to $\mathfrak{C}(\tilde{K}, \tilde{\Gamma})$. The local algebras for an arbitrary region O are then defined as

$$\mathfrak{A}(O) = \{\psi([k]) \mid k \in \tilde{K}, \mathrm{supp}(k) \subset O\}^{\mathrm{even}} , \tag{2.42}$$

where "even" refers to the subalgebra of even elements w.r.t. the $\mathbb{Z}_2$-grading of the CAR algebra.

We now describe the construction of the ground state representation on an ultra-static spacetime $\mathcal{M} = \mathbb{R} \times C$, $g = -dt^2 + h$ with compact Cauchy surface C, assumed to be spin. Let us assume that the orthonormal frame is chosen in such a manner that e_0 is the time-like normal to C. We may write the Dirac equation $(\nabla\!\!\!\!/ + m)\psi = 0$ as

$$i\partial_t \psi = \sum_{j=1}^{d} (ie_0 \cdot e_j \cdot \nabla_{e_j} + ie_0 m)\psi \equiv h\psi . \tag{2.43}$$

The right side defines a "1-particle Hamiltonian". By standard theorems, it has a discrete spectrum of eigen-spinors. It is checked that $\Gamma^{-1}h\Gamma = -h$. So, Γ exchanges eigen-spinors with positive and negative eigenvalues, and the spectrum is symmetric about 0. Around 0, there is a gap including at least $(-m, m)$. We let P be the projector in K onto the subspace of eigenspinors with positive eigenvalues, which as a consequence satisfies $\Gamma P \Gamma = 1 - P$. Thus, we can define, as in our general discussion of the CAR algebra in Sect. 2.2.2, a representation of the CAR algebra $\mathfrak{C}(K, \Gamma)$ associated with P on a fermionic Fock space. The 1-particle Hilbert space is $\mathcal{H}_1 = PK$. On the subspace $\mathcal{H}_n = E_n \mathcal{H}_1^{\otimes n}$ of "n-particle" states in fermonic Fock space, the Hamiltonian of the QFT is then given by

$$H|\Psi_n\rangle = \sum_{i=1}^{n} (1 \otimes \ldots \underbrace{h}_{i-\text{th slot}} \otimes \ldots 1)|\Psi_n\rangle , \qquad |\Psi_n\rangle \in \mathcal{H}_n . \qquad (2.44)$$

The vacuum state satisfies $H|0\rangle = 0$, so it is a ground state. The fermionic field operator is represented by (2.23).

Rather than unraveling these abstract definitions, we consider as a simple example the Majorana field in 1+1 dimensional Minkowski spacetime $\mathbb{R}^{1,1}$. In this example, $K = L^2(\mathbb{R}, dx; \mathbb{C}^2)$, Eq. (2.38) becomes the standard inner product on this space, and Γ becomes component-wise complex conjugation (using a suitable representation of the Clifford algebra). The 1-particle Hamiltonian becomes in this representation

$$h = \begin{pmatrix} p & im \\ -im & -p \end{pmatrix} \qquad (2.45)$$

with $p = id/dx$. The projector P onto the positive part of the spectrum can be worked out explicitly by diagonalizing this matrix. Once this has been done, it is convenient to identify the 1-particle Hilbert space PK with $L^2(\mathbb{R}, d\theta)$ via the isometry

$$V : PL^2(\mathbb{R}, dx; \mathbb{C}^2) \to L^2(\mathbb{R}, d\theta) =: \mathcal{H}_1 \qquad (2.46)$$

defined by

$$V \begin{pmatrix} k_1 \\ k_2 \end{pmatrix} = \frac{1}{\sqrt{2}} [e^{\theta/2 - i\pi/4} \tilde{k}_1(m \sinh \theta) + e^{-\theta/2 + i\pi/4} \tilde{k}_2(m \sinh \theta)] , \qquad (2.47)$$

where a tilde means Fourier transform. The Dirac field operator in the representation π_P (defined abstractly by (2.23)) becomes under this identification, and under the identification of K and $\bar{K}$ described in Sect. 2.2.2, the 2-component operator valued distribution

$$\psi(x) = \frac{1}{\sqrt{4\pi}} \int_{\mathbb{R}} d\theta \left\{ \begin{pmatrix} e^{\theta/2 - i\pi/4} \\ e^{-\theta/2 + i\pi/4} \end{pmatrix} e^{-ip(\theta)x} a^{\dagger}(\theta) + \begin{pmatrix} e^{\theta/2 + i\pi/4} \\ e^{-\theta/2 - i\pi/4} \end{pmatrix} e^{ip(\theta)x} a(\theta) \right\} , \qquad (2.48)$$

where $x \equiv (x^0, x^1)$, where $(p_\mu(\theta)) = (-m \cosh \theta, m \sinh \theta)$, and where $a(\theta), a^{\dagger}(\theta)$ satisfy the relations (2.51) below with $S_2 = -1$. A wave function $|\Phi_1\rangle \in \mathcal{H}_1$ is created from the vacuum by applying a smeared creation operator, $|\Phi_1\rangle = \int d\theta \phi(\theta) a^{\dagger}(\theta)|0\rangle \equiv a(\phi)^* |0\rangle$, with $\phi \in L^2(\mathbb{R}, d\theta)$, compare Sect. 2.2.2.

2.4.3 Integrable Models in 1+1 Dimension with Factorizing S-Matrix

Following an idea of [34], it has been shown in [35–37], how to construct a wide class of integrable models in 1+1 dimensional Minkowski space satisfying the above axioms (a1)–(a4) and a nuclearity condition [(a5) and (a5')]. The only input[11] in this construction is a 2-body scattering-matrix. We will not discuss here how in general the concept of a scattering-matrix fits into the algebraic framework, see [38] for a discussion and references. For the sake of the construction, it is enough to think of the 2-body scattering-matrix as merely a function S_2 which is a datum entering the construction of a net $\mathfrak{A}(O)$. The properties required of S_2 to make this construction work are:

(s1) $S_2(\theta)$ is a bounded analytic function on the strip $\{\theta \in \mathbb{C} \mid -\epsilon < \Im\theta < \pi + \epsilon\}$ where $0 < \epsilon < \pi/2$.

(s2) $|S_2(\theta)| = 1$ for $\theta \in \mathbb{R}$, and $S_2(0) = -1$ (it is therefore "fermionic" in the terminology of [35]).

(s3) For θ in $\mathbb{R}$, $S_2(-\theta) = S_2(\theta)^{-1}$.

(s4) For θ in $\mathbb{R}$, $S_2(\theta + i\pi) = S_2(\theta)^{-1}$.

(s1) corresponds to analyticity, (s2) to unitarity, (s3) to PCT-invariance, and (s4) to crossing symmetry for the full scattering-matrix, if $k = (m \cosh\theta, m \sinh\theta)$ is identified with the incoming momentum of an on-shell particle of mass m in a 2-body collision. For instance, in the Sinh-Gordon model [39]

$$S_2^{\mathrm{ShG}}(\theta) = \frac{\sinh\theta - i \sin b}{\sinh\theta + i \sin b}, \quad b = \frac{\pi g^2}{1 + g^2} \tag{2.49}$$

with g the coupling constant of the Sinh-Gordon potential. S_2^{ShG} satisfies (s1)–(s4) as long as $0 < b < \pi$, as would for instance any product of an odd number of such factors with different $0 < b_i < \pi$.

The construction of the net $O \mapsto \mathfrak{A}(O)$ corresponding to a given S_2 starts by considering an "S_2-symmetric" Fock-space over $\mathcal{H}_1 = L^2(\mathbb{R}, d\theta)$. This Fock space is a direct sum $\mathbb{C} \oplus_{n \geq 1} \mathcal{H}_n$ of n-particle spaces. By contrast to the case of the bosonic Fock-space, $\mathcal{H}_n$ is not obtained by applying a symmetrization projection to $\mathcal{H}_1^{\otimes n}$. Rather, one applies a projection E_n based on S_2. For that, let τ_i be an elementary transposition of the elements i with $i + 1$ in the symmetric group $\mathfrak{S}_n$ on n elements. Define an exchange operator $D_n(\tau_i)$ on $\mathcal{H}_1^{\otimes n}$, identified with unsymmetrized L^2-wave functions Ψ_n in n-variables, as

$$(D_n(\tau_i)\Psi_n)(\theta_1, \ldots, \theta_i, \theta_{i+1}, \ldots, \theta_n) = S_2(\theta_{i+1} - \theta_i)\Psi_n(\theta_1, \ldots, \theta_{i+1}, \theta_i, \ldots, \theta_n) . \tag{2.50}$$

---

[11]Apart from the value m of the mass of the basic particle.

It can be shown using (s2)–(s3) that this exchange operator gives a unitary representation[12] of $\mathfrak{S}_n$ on $\mathcal{H}_1^{\otimes n}$. Define an S_2-symmetric projection $E_n = (1/n!)$ $\sum_{\sigma \in \mathfrak{S}_n} D_n(\sigma)$, define $\mathcal{H}_n = E_n \mathcal{H}_1^{\otimes n}$ (S_2-symmetric wave functions), define $\mathcal{H} = \mathbb{C} \oplus_{n \geq 1} \mathcal{H}_n$ and define creation and annihilation operators $z^\dagger(\Psi)$, $z(\Psi)$, $\Psi \in \mathcal{H}_1$ on $\mathcal{H}$ by analogy with Eq. (2.17). We write informally[13] $z^\dagger(\Psi) = \int d\theta \Psi(\theta) z^\dagger(\theta)$. These operators satisfy relations called the "Zamolodchikov-Faddeev (ZF) algebra", which is

$$z(\theta) z^\dagger(\theta') - S_2(\theta - \theta') z^\dagger(\theta') z(\theta) = \delta(\theta - \theta') \cdot 1, \quad z(\theta) z(\theta') - S_2(\theta' - \theta) z(\theta') z(\theta) = 0. \quad (2.51)$$

Following Schroer and Wiesbrock [34], one then defines a "field operator" by putting, with $px = p_\mu x^\mu$ and $(p_\mu(\theta)) = (-m \cosh \theta, m \sinh \theta)$,

$$\phi(x) = \frac{1}{\sqrt{4\pi}} \int_{\mathbb{R}} d\theta [z^\dagger(\theta) e^{-ip(\theta)x} + z(\theta) e^{+ip(\theta)x}]. \quad (2.52)$$

This field satisfies the KG equation with mass m. Its linear structure in creation- and annihilation operators is analogous to that of a KG quantum field in 1+1 dimensions in the vacuum representation (see (2.36)), when we identify $z^\dagger(\theta) = \frac{1}{\sqrt{2}} a^\dagger(k_1(\theta))$. Actually, for $S_2 = 1$, $\phi(x)$ is exactly equal to the free KG field because the ZF generators then satisfy the standard relations of creation and annihilation operators. In this special case, the field $\phi(x)$ defined by (2.52) satisfies space-like commutativity. But in general, it does not have this property. It may, however, be used to define a local quantum field theory in the sense of axioms (a1)–(a4) by a roundabout route. First, define the following anti-linear operator J_n on $\mathcal{H}_1^{\otimes n}$:

$$(J_n \Psi_n)(\theta_1, \ldots, \theta_n) = \overline{\Psi_n(\theta_n, \ldots, \theta_1)}. \quad (2.53)$$

The properties of S_2 imply that this consistently defines, in fact, an anti-unitary, involutive, operator on the S_2-symmetric Fock space $\mathcal{H}$ (it commutes with E_n). Call this operator[14] J, and define $z'(\theta) = J z(\theta) J$, $z'^\dagger(\theta) = J z^\dagger(\theta) J$. These operators satisfy relations that are identical to the ZF-algebra, except that $S_2(\theta - \theta')$ is replaced by $S_2(\theta' - \theta)$. Define a field $\phi'(x)$ by substituting into (2.52) the primed ZF creation and annihilation operators. The key observation [34], which follows from (s2)–(s4), is:

Lemma 1 *Assume f, f' are test functions on $\mathbb{R}^2$ such that the support of f' is space-like and to the left of that of f. Then $[\phi(f), \phi'(f')] = 0$ (on a core of vectors in $\mathcal{H}$).*

One also shows using Nelson's analytic vector theorem [31] (just as in the case of the free KG-field) that the operators $\phi(f)$, $\phi'(f')$ are (closable and) essentially

[12] I.e., it satisfies the relations of the permutation group.

[13] Informally, $z^\dagger(\theta) = z(\theta)^*$.

[14] J turns out to be equal to the modular conjugation associated with the algebra $\mathfrak{R}$ defined below.

self-adjoint. Then their exponentials are well-defined, and the lemma holds also for them, i.e. $[e^{i\phi(f)}, e^{i\phi'(f')}] = 0$. Let $W = \{(t, x) \in \mathbb{R}^2 \mid x > |t|\}$ be the right wedge in $\mathbb{R}^2$. Based on the "half-sided" locality expressed by the lemma, it is natural to define the wedge algebras

$$\mathfrak{R} = \{e^{i\phi(f)} \mid \text{supp}(f) \subset W\}'', \quad \mathfrak{R}' = \{e^{i\phi'(f')} \mid \text{supp}(f') \subset W'\}'', \qquad (2.54)$$

where W' is the opposite wedge and the double prime is the v. Neumann closure. As the notation suggests, the commutation relations of the lemma not only imply that $\mathfrak{R}'$ commutes with $\mathfrak{R}$, but even that $\mathfrak{R}'$ is in fact the commutant, i.e. the set of all such operators. To get algebras associated with bounded double cones, one would like to form intersections of appropriately "translated" wedge algebras. For this, we first need a representation of translations (and Lorentz-boosts) on $\mathcal{H}$. These are defined by the unitaries ($a \in \mathbb{R}^2, \lambda \in \mathbb{R}$)

$$(U_n(\lambda, a)\Psi_n)(\theta_1, \ldots, \theta_n) = \exp\left(-i\sum_j p(\theta_j)a\right)\Psi_n(\theta_1 - \lambda, \ldots, \theta_n - \lambda),$$

$$(2.55)$$

where λ is interpreted as the boost-parameter and a as the translation vector of an element $g = (\lambda, a)$ of the 2-dimensional Poincaré-group. The properties of S_2 once again imply that this consistently defines a strongly-continuous, positive energy representation U of P on $\mathcal{H}$. If W_a denotes the translation of W by the vector $a \in \mathbb{R}^2$, then any double cone O can be written as an intersection of opposite, translated wedges, $O = W_a \cap W_b'$, for suitable $a, b \in \mathbb{R}^2$. Thus, it is natural to define:

Definition 6 Let $O = W_a \cap W_b'$, for suitable $a, b \in \mathbb{R}^2$. Then we define a net by

$$O \mapsto \mathfrak{A}(O) \equiv U(a)\mathfrak{R}U(a)^* \cap U(b)\mathfrak{R}'U(b)^*. \qquad (2.56)$$

It is then straightforward to see that this net $O \mapsto \mathfrak{A}(O)$ satisfies axioms (a1)–(a5) (for v. Neumann algebras), where the vacuum state is just that in the S_2-symmetric Fock space. It is not so clear, however, that this net is non-trivial, but this has been established for sufficiently large regions[15] O in [35, 40] using a combination of techniques such as modular nuclearity bounds, analyticity methods, and making heavy use of the properties of S_2. The methods also show that each $\mathfrak{A}(O)$ is of type III$_1$, and that the vacuum vector $|0\rangle$ in our Fock space $\mathcal{H}$ is cyclic and separating.

The surprisingly simple, in principle, construction of a wide class of integrable models just outlined has one serious caveat, though. Unlike for the case of the free KG-field, the operator $\phi(x)$ (and likewise $\phi'(x)$) is *not* a local operator. In particular, for instance in the Sinh-Gordon model, $\phi(x)$ must not be interpreted as a quantum version of the classical field appearing in the corresponding Lagrangian (1.14), because that would be expected to be local. Local operators at a point x can be characterized

---

[15]Note that [35] contained an error, which has been amended in [40]. At present, the arguments only establish (a1)–(a4), (a5') for regions O of a minimal size, contrary to the claim of [35].

in principle as being, in a certain sense [41, 42], elements in $\cap_{O \ni x} \mathfrak{A}(O)$. But in practice, they would in all likelihood have an extremely complicated expression in terms of the non-local operators ϕ and ϕ', and in this sense one can only say that the model has only been constructed in a very indirect way. Fortunately, for our purposes, it will not be important at all to precisely identify such local operators. Rather, all we need is the information given by the net $\{\mathfrak{A}(O)\}$, which has a straightforward and simple definition.

2.4.4 Chiral CFTs

Chiral conformal field theories (CFTs) describe "one chiral half" of a conformal field theory in 1+1 dimensions, and are particularly well-investigated. They are described in the algebraic setting by nets $I \mapsto \mathfrak{A}(I)$ parameterized by open intervals of the circle $I \subset S^1$. The axioms are essentially the same as in Minkowski space, with a few fairly evident changes. In the commutativity axiom (a2), one replaces the notion of causal complement simply by disjoint intervals, i.e. $[\mathfrak{A}(I_1), \mathfrak{A}(I_2)] = \{0\}$ if I_1 and I_2 are disjoint. The covariance axiom now involves the Möbius group $G = SU(1, 1)/\{1, -1\} = PSU(1, 1)$, rather than the Poincaré group. An element $g \in PSU(1, 1)$ acts on $z \in S^1$ by

$$g \cdot z = \frac{\alpha z + \beta}{\bar{\beta} z + \bar{\alpha}}, \qquad g = \begin{pmatrix} \alpha & \beta \\ \bar{\beta} & \bar{\alpha} \end{pmatrix} \quad \text{where } |\alpha|^2 - |\beta|^2 = 1 . \tag{2.57}$$

In the transcription of the covariance and vacuum requirements (a3) and (a4), one requires a unitary, positive energy representation U of the group G (or its cover $\widetilde{G}$) replacing the group of "spacetime" symmetries. A net $\{\mathfrak{A}(I)\}$ over S^1 with these properties is called a "chiral net".[16]

Under the Cayley transform $z \mapsto i(z - 1)/(z + 1)$, the circle gets mapped to the 1-point compactification of $\mathbb{R}$ and the action of the Möbius group then corresponds to the action of $G \cong PSL(2, \mathbb{R})$ by the transformations $\mathbb{R} \ni x \mapsto g \cdot x = \frac{ax+b}{cx+d}$ with real coefficients such that $ad - bc = 1$. This action is just the action of the conformal group $SO_+(2, 2) \cong G \times G$ on light rays of 2-dimensional Minkowski space. A corresponding 2-dimensional conformal net in 1+1 dimensions can therefore be defined simply as follows. Viewing 2-dimensional Minkowski spacetime as the Cartesian product of two light rays, any double cone O is the Cartesian product of two open intervals $I_L, I_R \subset \mathbb{R}$. We then simply set $\mathfrak{A}_{2d}(O) = \mathfrak{A}(I_L) \otimes \mathfrak{A}(I_R)$ for $O = I_L \times I_R$. Such a net on $\mathbb{R}^2$ is thus the tensor product of chiral theories. It again satisfies analogous versions of the axioms (a1)–(a4), with the group of space-

---

[16]One sometimes requires that the symmetry algebra of the net is the full Virasoro algebra, i.e. that the net contains the algebra of quantized diffeomorphisms as a subnet. Then the split property is automatic [43].

time symmetries now replaced by conformal transformations or its cover $\widetilde{\mathrm{SO}}_+(2,2)$. (More details can be found e.g. in [44, 45].)

There are very many examples for chiral nets, for a detailed discussion in the case of a central charge $c < 1$ see e.g. [46]. Here we only give one such example purely for illustrative purposes. It is an operator algebraic version of the so-called "minimal models," see e.g. [47] for the conventional description. A local algebra $\mathfrak{A}(I)$ in such a net basically describes "quantized diffeomorphisms" on S^1 acting non-trivially only in the interval $I \subset S^1$. The construction is more precisely as follows, see e.g. [48] for details and references.

Any diffeomorphism $\tilde{f}$ of $\mathbb{R}$ satisfying $\tilde{f}(\theta + 2\pi) = \tilde{f}(\theta) + 2\pi$ defines an orientation preserving diffeomorphism f of S^1 via the formula $f(e^{i\theta}) = e^{i\tilde{f}(\theta)}$. Call the group of these $\mathrm{Diff}_+(S^1)$. The "Bott cocycle" is the map $B : \mathrm{Diff}_+(S^1) \times \mathrm{Diff}_+(S^1) \to \mathbb{R}$ defined by

$$B(f_1, f_2) = -\frac{1}{48\pi} \int_{S^1} \ln((f_1 \circ f_2)'(z)) \frac{\mathrm{d}}{\mathrm{d}z} \ln(f_2'(z)) \, \mathrm{d}z. \qquad (2.58)$$

It can be shown that this lifts to a cocycle of the universal covering group $\widetilde{\mathrm{Diff}}_+(S^1)$, which is shown to be a Fréchet Lie-group. A unitary representation U of this group on some Hilbert space $\mathcal{H}$ is called a multiplier representation with central charge $c \in \mathbb{R}$ if it is strongly continuous with respect to the group topology and

$$U(f_1)U(f_2) = e^{icB(f_1, f_2)} U(f_1 \circ f_2) . \qquad (2.59)$$

The universal covering $\widetilde{G}$ of the Möbius group is a subgroup of $\widetilde{\mathrm{Diff}}_+(S^1)$. For $f_1, f_2 \in \widetilde{G}$, the Bott-cocycle vanishes, so U restricts to a bona-fide, ordinary, unitary representation of $\widetilde{G}$. For given c, the irreducible multiplier representations can be classified. They correspond to exponentiated versions of the "highest weight representations" of the Virasoro-algebra, which is an infinitesimal version of the relation (2.59), the Bott cocycle B corresponding to the central term in the Virasoro-algebra. Such representations only exist for $c > 0$, and for $c < 1$, the central charge must be quantized according to the rule $c = 1 - \frac{6(p-p')^2}{pp'}$, $p, p' \in \mathbb{N}$. The vacuum representation U_0 corresponds to the highest weight representation in which the generator L_0^{vac} of rotations of S^1 satisfies $L_0^{\mathrm{vac}}|0\rangle = 0$, where $|0\rangle \in \mathcal{H}_0$ is the highest weight vector (vacuum). The net $I \subset S^1 \mapsto \mathfrak{A}(I)$ is defined by

$$\mathfrak{A}(I) = \{U_0(f) \mid f \in \widetilde{\mathrm{Diff}}_+(S^1) \text{ such that } f(z) = z \text{ for } z \notin I\}'' . \qquad (2.60)$$

Using properties of multiplier representations [49], one shows that this definition satisfies the analogues of axioms (a1)–(a4) for chiral CFTs.

References

1. R.V. Kadison, J.R. Ringrose, *Fundamentals of the Theory of Operator Algebras* (Academic Press, New York, I 1983, II 1986)
2. O. Bratteli, D.W. Robinson, *Operator Algebras and Quantum Statistical Mechanics* (Springer, I 1987, II 1997)
3. R.T. Powers, E. Størmer, Free states of the canonical anticommutation relations. Commun. Math. Phys. **16**, 1–33 (1970)
4. H. Araki, Relative entropy for states of von Neumann algebras. Publ. RIMS Kyoto Univ. **11**, 809–833 (1976)
5. H. Araki, Relative entropy for states of von Neumann algebras II. Publ. RIMS Kyoto Univ. **13**, 173–192 (1977)
6. M. Florig, S.J. Summers, On the statistical independence of algebras of observables. J. Math. Phys. **38**, 1318 (1997)
7. S. Doplicher, R. Longo, Standard and split inclusions of von Neumann algebras. Invent. Math. **75**, 493 (1984)
8. D. Buchholz, E.H. Wichmann, Causal independence and the energy level density of states in local quantum field theory. Commun. Math. Phys. **106**, 321 (1986)
9. E. Binz, R. Honegger, A. Rieckers, Construction and uniqueness of the C^*-Weyl algebra over a general pre-symplectic space. J. Math. Phys. **45**, 2885–2907 (2004)
10. F. Kärsten, *Report MATH, 89–06* (Akademie der Wissenschaften der DDR, Karl-Weierstrass-Institut für Mathematik, Berlin, 1989)
11. R.M. Wald, *Quantum Field Theory in Curved Space-Time and Black Hole Thermodynamics* (Chicago University Press, Chicago, 1994)
12. J. Manuceau, A. Verbeure, Quasi-free states of the C.C.R.-algebra and Bogoliubov transformations. Commun. Math. Phys. **9**, 293–302 (1968)
13. H. Araki, S. Yamagami, On quasi-equivalence of quasifree states of the canonical commutation relations. Publ. RIMS Kyoto Univ. **18**, 283–338 (1982)
14. P. Leylands, J.E. Roberts, D. Testard, *Duality for Quantum Free Fields* (Preprint CNRS, Marseille, 1978)
15. H. Araki, On quasifree states of the CAR and Bogoliubov automorphisms. Publ. RIMS Kyoto Univ. **6**, 385–442 (1970)
16. J. Cuntz, Simple C*-algebras generated by isometries. Commun. Math. Phys. **57**, 173–185 (1977)
17. S. Hollands, R.M. Wald, Quantum fields in curved spacetime. Phys. Rept. **574**, 1 (2015)
18. H. Araki, *Mathematical Theory of Quantum Fields* (Oxford Science Publications, 1993)
19. H. Reeh, S. Schlieder, Bemerkungen zur Unitäräquivalenz von lorentzinvarianten Feldern. Nuovo Cimento **22**, 1051–1068 (1961)
20. R. Haag, J.A. Swieca, When does a quantum field theory describe particles? Commun. Math. Phys. **1**, 308–320 (1965)
21. A. Pietsch, *Nuclear Locally Convex Spaces* (Springer, Berlin, 1972)
22. C.J. Fewster, I. Ojima, M. Porrmann, p-nuclearity in a new perspective. Lett. Math. Phys. **73**, 1–15 (2005)
23. D. Buchholz, K. Fredenhagen, C. D'Antoni, The universal structure of local algebras. Commun. Math. Phys. **111**, 123 (1987)
24. C.J. Fewster, R. Verch, The necessity of the Hadamard condition. Class. Quant. Grav. **30**, 235027 (2013)
25. D. Buchholz, C. D'Antoni, R. Longo, Nuclear maps and modular structures. 1. General properties. J. Funct. Anal. **88**, 223 (1990)
26. G. Lechner, K. Sanders, Modular nuclearity: a generally covariant perspective. Axioms **5**, 5 (2016)
27. D. Buchholz, P. Jacobi, On the nuclearity condition for massless fields. Lett. Math. Phys. **13**, 313 (1987)

28. C. Bär, N. Ginoux, F. Pfäffle, *Wave Equations on Lorentzian Manifolds and Quantization*, ESI Lectures in Mathematics and Physics (European Mathematical Society Publishing House, Germany, 2007)
29. R. Seeley, Complex powers of elliptic operators. Proc. Symp. Pure Appl. Math. AMS **10**, 288–307 (1967)
30. B.S. Kay, Linear spin-zero quantum fields in external gravitational and scalar fields I. A one particle structure for the stationary case. Commun. Math. Phys. **62**, 55–70 (1978)
31. M. Reed, B. Simon, *Methods of Modern Mathematical Physics II* (Academic Press, 1975)
32. C. D'Antoni, S. Hollands, Nuclearity, local quasiequivalence and split property for Dirac quantum fields in curved space-time. Commun. Math. Phys. **261**, 133 (2006)
33. J. Figueroa-O'Farrill, *Majorana Spinors*, http://www.maths.ed.ac.uk/jmf/Teaching/Lectures/Majorana.pdf
34. B. Schroer, H.W. Wiesbrock, Modular constructions of quantum field theories with interactions. Rev. Math. Phys. **12**, 301 (2000)
35. G. Lechner, Construction of quantum field theories with factorizing s-matrices. Commun. Math. Phys. **277**, 821 (2008)
36. G. Lechner, On the existence of local observables in theories with a factorizing S-matrix. J. Phys. A **38**, 3045–3056 (2005)
37. D. Buchholz, G. Lechner, Modular nuclearity and localization. Ann. Henri Poincaré **5**, 1065 (2004)
38. R. Haag, *Local Quantum Physics: Fields, Particles, Algebras* (Springer, Berlin, 1992)
39. E. Abdalla, C. Abdalla, K.D. Rothe, *Non-perturbative Methods in 2-Dimensional Quantum Field Theory* (World Scientific, New Jersey, 1991)
40. S. Alazzawi, G. Lechner, *Inverse Scattering and Locality in Integrable Quantum Field Theories*, arXiv:1608.02359 [math-ph]
41. K. Fredenhagen, J. Hertel, Local algebras of observables and point-like localized fields. Commun. Math. Phys. **80**, 555 (1981)
42. H. Bostelmann, Phase space properties and the short distance structure in quantum field theory. J. Math. Phys. **4**, 052301 (2005)
43. V. Morinelli, Y. Tanimoto, M. Weiner, *Conformal Covariance and the Split Property*, Commun. Math. Phys. **357**(1), 379-406 (2018)
44. Y. Kawahigashi, Conformal field theory, tensor categories and operator algebras. J. Phys. A **48**(30), 303001, 57 (2015)
45. K.-H. Rehren, Algebraic conformal quantum field theory in perspective, in *Advances in Algebraic Quantum Field Theory* (Springer, Switzerland, 2015), pp. 331–364. Math. Phys. Stud
46. Y. Kawahigashi, R. Longo, Classification of local conformal nets: case c < 1. Ann. Math. **160**, 493 (2004)
47. P. DiFrancesco, P. Mathieu, D. Senechal, *Conformal Field Theory* (Springer, New York, 1997)
48. C.J. Fewster, S. Hollands, Quantum energy inequalities in two-dimensional conformal field theory. Rev. Math. Phys. **17**, 577 (2005)
49. V.T. Laredo, Integrating unitary representations of infinite-dimensional Lie groups. J. Funct. Anal. **161**, 478–508 (1999)

Chapter 3
Entanglement Measures in QFT

Abstract In this chapter we discuss entanglement in a general setting and we review some quantitative measures of entanglement and their properties.

In this section we discuss entanglement in a general setting and we review some quantitative measures of entanglement and their properties. (See [1–4] for more details.)

3.1 Entanglement

Let us begin by introducing the basic notion of entanglement. We consider a system, described by a C^* (or v. Neumann)-algebra $\mathfrak{A}$, and two subsystems described by subalgebras $\mathfrak{A}_A, \mathfrak{A}_B \subset \mathfrak{A}$. Furthermore, we let ω be a state on $\mathfrak{A}$ and we wish to characterize and quantify the entanglement in the state ω between the algebras $\mathfrak{A}_A$ and $\mathfrak{A}_B$.

Using the GNS-representation π_ω on $\mathcal{H}_\omega$, we may replace the C^*-algebras by the corresponding v. Neumann algebras, which we will again denote by $\mathfrak{A}, \mathfrak{A}_A$ and $\mathfrak{A}_B$, respectively. (Note, however, that these algebras may depend in general on the choice of ω.) We will assume that $\mathfrak{A}_A$ and $\mathfrak{A}_B$ commute, that $\mathfrak{A}_A \cap \mathfrak{A}_B = \mathbb{C}1$, and that $\mathfrak{A}_A$ and $\mathfrak{A}_B$ are statistically independent, i.e. $\mathfrak{A}_A \vee \mathfrak{A}_B \simeq \mathfrak{A}_A \otimes \mathfrak{A}_B$, where the latter algebra acts on $\mathcal{H}_\omega \otimes \mathcal{H}_\omega$. Since ω restricts to a normal state on $\mathfrak{A}_A \vee \mathfrak{A}_B$, we can view this restriction also as a normal state on $\mathfrak{A}_A \otimes \mathfrak{A}_B$, which is described by a density matrix ρ on $\mathcal{H}_\omega \otimes \mathcal{H}_\omega$.

We can now introduce the distinction between "separable states" and "entangled states":

Definition 7 A normal state ω on the tensor product $\mathfrak{A}_A \otimes \mathfrak{A}_B$ of two v. Neumann algebras is said to be "separable" if it can be written as a norm convergent sum

© The Author(s), under exclusive licence to Springer Nature Switzerland AG 2018   43
S. Hollands and K. Sanders, *Entanglement Measures and Their Properties in Quantum Field Theory*, SpringerBriefs in Mathematical Physics 34,
https://doi.org/10.1007/978-3-319-94902-4_3

$\omega = \sum_j \varphi_j \otimes \psi_j$ for positive normal functionals φ_j, ψ_j on $\mathfrak{A}_A$ respectively $\mathfrak{A}_B$, i.e. $\omega(ab) = \sum_j \varphi_j(a)\psi_j(b)$. A normal state which is not separable is called "entangled".

When $\mathfrak{A}_A \otimes \mathfrak{A}_B$ is finite dimensional, the set of separable states is norm-closed.[1]

The simplest example of a separable state is $\omega = \omega_A \otimes \omega_B$ for two vector states $\omega_A(a) = \langle \Phi | a\Phi \rangle$ and $\omega_A(b) = \langle \Psi | b\Psi \rangle$. By definition, $\omega(ab) = \omega_A(a)\omega_B(b)$, which is a vector state determined by the simple tensor product vector $|\Phi\rangle \otimes |\Psi\rangle$. Alternatively, we can write the states ω_A and ω_B in terms of the density matrices ρ_A and ρ_B, which are simply orthogonal projections onto $|\Phi\rangle$ and $|\Psi\rangle$, respectively. The state ω is then determined by the density matrix $\rho = \rho_A \otimes \rho_B$. We remind the reader that in our general setting, vector states need not be pure and density matrices need not be uniquely determined by the state ω.

A general separable state is always a convex combination of such separable vector states. Indeed, for a general separable state ω, decomposed in terms of φ_j and ψ_j, we have $\varphi_j(a) = \mathrm{Tr}(\rho_{A,j}a)$ and $\psi_j(b) = \mathrm{Tr}(\rho_{B,j}b)$ for suitable positive trace-class operators $\rho_{A,j}, \rho_{B,j}$. We can then write the density matrix as $\rho = \sum_j \rho_{A,j} \otimes \rho_{B,j}$. By diagonalising the operators $\rho_{A,j}$ and $\rho_{B,j}$, decomposing them into one-dimensional projectors, and relabelling indices we can always write ρ as a sum of one-dimensional projectors which project onto vectors of simple tensor product form. Hence, ω is a convex combination of separable vector states.

Note that a general unit vector $\Theta \in \mathcal{H}_A \otimes \mathcal{H}_B$ can be written as a sum of simple tensor products, $\Theta = \sum_j |\Phi_j\rangle \otimes |\Psi_j\rangle$, but the corresponding positive linear functional is $\langle \Theta, ab\Theta \rangle = \sum_{j,k} \langle \Phi_j | a\Phi_k \rangle \langle \Psi_j | b\Psi_k \rangle$, which need not be separable. Analogously, the density matrix ρ of any normal state ω on $\mathfrak{A}_A \otimes \mathfrak{A}_B$ can be diagonalised and then decomposed as $\rho = \sum_j x_j \otimes y_j$ with bounded operators x_j, y_j, but in general these operators cannot be chosen positive (although they can be chosen self-adjoint). The obstruction is what characterizes entanglement. Another reformulation of this obstruction is that any normal state ω on $\mathfrak{A}_A \otimes \mathfrak{A}_B$ can be written as $\omega = \sum_j \varphi_j \otimes \psi_j$ with hermitean normal functionals φ_j, ψ_j which may not be positive.

When $\mathfrak{A}_A$ and $\mathfrak{A}_B$ are in standard form with vectors $|\Omega_A\rangle$ and $|\Omega_B\rangle$, then $\mathfrak{A}_A \otimes \mathfrak{A}_B$ is also in standard form with vector $|\Omega\rangle := |\Omega_A\rangle \otimes |\Omega_B\rangle$. Recall that every normal state ω has a unique vector representative in the natural cone $\mathcal{P}^\sharp \subset \mathcal{H} = \mathcal{H}_A \otimes \mathcal{H}_B$. If $\mathcal{P}_A^\sharp, \mathcal{P}_B^\sharp$ denotes the natural cone in $\mathcal{H}_A, \mathcal{H}_B$ respectively, and if $|\Psi_A\rangle \in \mathcal{P}_A^\sharp$ and $|\Psi_B\rangle \in \mathcal{P}_B^\sharp$ are unit vectors, then $|\Psi_A\rangle \otimes |\Psi_B\rangle \in \mathcal{P}^\sharp$ is a separable state. All pure separable states are necessarily of this form, and using the properties of the natural cones one may show that the norm limit of separable states of the form $\psi \otimes \phi$ is again separable. Note, however, that it is not so easy to recognize when a vector in $\mathcal{P}^\sharp$ defines a mixed separable state, because the separable states do not form a cone inside $\mathcal{P}^\sharp$.

Depending on the state ω, the outcomes of separate measurements on the two systems $\mathfrak{A}_A$ and $\mathfrak{A}_B$ can exhibit different kinds of correlations. When $\omega = \omega_A \otimes \omega_B$,

---

[1]The general case is unclear to us, but one could modify the definition to make the set of separable states norm-closed.

there are no correlations at all. For a general separable state, however, there can
be correlations, which are of a classical nature. Entangled states exhibit even more
general "quantum" correlations. For this reason, entanglement has come to be viewed
as an experimental resource, which can be enhanced or "purified", and subsequently
exploited to perform quantum computations, teleportations or other often counter-
intuitive experiments.

3.2 Properties of Entanglement Measures

Let us now turn to the question how to quantify the amount of entanglement in a
general normal state ω on $\mathfrak{A}_A \otimes \mathfrak{A}_B$. We will start in this section by reviewing a
number of desirable properties that an entanglement measure $E(\omega)$ could satisfy. In
the remainder of this chapter we will then introduce specific examples and discuss
the properties that they have.

We start with the following basic properties:

(e0) (symmetry) $E(\omega)$ is independent of the order of the systems A and B.
(e1) (non-negative) $E(\omega) \in [0, \infty]$, with $E(\omega) = 0$ if and only if ω is separable,
   and $E(\omega) = \infty$ when ω is not a normal state on $\mathfrak{A}_A \otimes \mathfrak{A}_B$ (e.g. when $\mathfrak{A}_A$ and
   $\mathfrak{A}_B$ are not statistically independent).
(e2) (continuity) Let $\mathfrak{N}_{A1} \subset \mathfrak{N}_{A2} \cdots \subset \mathfrak{N}_{Ai} \cdots \subset \mathfrak{A}_A$ be an increasing net of type
   I factors isomorphic to matrix algebras $\mathfrak{N}_i \cong M_{n_i}(\mathbb{C})$, and similarly for B. Let
   ω_i, ω_i' be normal states on $\mathfrak{N}_{Ai} \otimes \mathfrak{N}_{Bi}$ such that $\lim_{i\to\infty} \|\omega_i' - \omega_i\| = 0$. Then

$$\lim_{i\to\infty} \frac{E(\omega_i') - E(\omega_i)}{\ln n_i} = 0. \tag{3.1}$$

(e3) (convexity) If $\omega = \sum_j \lambda_j \omega_j$ is a convex combination of states ω_j (with $\lambda_j \geq
   0, \sum_j \lambda_j = 1$), then

$$E(\omega) \leq \sum_j \lambda_j E(\omega_j), \tag{3.2}$$

   i.e. $\omega \mapsto E(\omega)$ is convex.

Property (e3) states that entanglement cannot be increased by mixing states. It can
be reduced, however: for two independent spin-$\frac{1}{2}$ systems, one can choose a Bell-
basis of four vectors in the tensor product Hilbert space. These vectors define pure,
(maximally) entangled states, but an equal mixture of these four states yields the
density matrix $\rho \propto 1 = 1_A \otimes 1_B$, which is separable.

The next property is based on the idea that certain experimental manipulations
cannot increase the amount of entanglement, because they can only introduce classical correlations between measurement results. Before we can formulate this property,
we will first review the allowed experimental operations, for which we will use the
following terminology:

Definition 8 A linear map $\mathcal{F} : \mathfrak{A}_2 \to \mathfrak{A}_1$ between two C^*-algebras is called **positive** (p) if $\mathcal{F}(a)$ is a positive operator whenever a is. $\mathcal{F}$ is called **completely positive** (cp) if $1_{M_N(\mathbb{C})} \otimes \mathcal{F}$ is positive as a map[2] $M_N(\mathbb{C}) \otimes \mathfrak{A}_2 \to M_N(\mathbb{C}) \otimes \mathfrak{A}_1$ for all N. A (completely) positive map is called **normalized** if $\mathcal{F}(1) = 1$. A normalized positive map $\mathcal{F}$ between v. Neumann algebras is called **normal** when $\mathcal{F}^*$ maps normal states to normal states.

A normalized positive map $\mathcal{F}$ gives rise to a map $\mathcal{F}^*$ from states on $\mathfrak{A}_1$ to states on $\mathfrak{A}_2$, defined by $(\mathcal{F}^*\omega)(a) := \omega(\mathcal{F}(a))$. (This point of view explains the order of $\mathfrak{A}_1$ and $\mathfrak{A}_2$ in the definition above.) Conversely, any map $\mathcal{F}^*$ from states on $\mathfrak{A}_1$ to states on $\mathfrak{A}_2$ arises from a normalized positive linear map in this way. Complete positivity is motivated by the desire to be able to apply the same experimental manipulations independently to N copies of the same system. It is the mathematical characterization of a "quantum channel" in the sense of Quantum Information Theory.

In addition to a quantum channel, one could perform measurements and post-select a sub-ensemble according to the results. For a v. Neumann measurement, given by projections $P_k \in \mathfrak{A}$ with $\sum_k P_k = 1$, we note that the maps $\mathcal{F}_k : \mathfrak{A} \to \mathfrak{A}$ defined by $a \mapsto P_k a P_k$ are cp, with $0 \leq \mathcal{F}_k(1) = P_k \leq 1$. Performing the measurement on a state ω we obtain the new state

$$\omega_k := \frac{\mathcal{F}_k^*\omega}{\omega(P_k)}$$

with probability $\omega(P_k)$, when $\omega(P_k) > 0$. A combination of quantum channels and measurements is called an "operation" [5]. It is described by a family $\mathcal{F}_k : \mathfrak{A}_2 \to \mathfrak{A}_1$ of cp maps with $\sum_k \mathcal{F}_k(1) = 1$, which transform a state ω on $\mathfrak{A}_1$ into $\omega_k := \frac{1}{p_k}\mathcal{F}_k^*\omega$ with probability $p_k := \omega(\mathcal{F}_k(1))$ when $p_k > 0$.

Example 1 Let us give some examples of p and cp maps.

   (i) Any (unit preserving) $*$-homomorphism between C^*-algebras (and in particular every representation) is a (normalized) cp map. Furthermore, any state of a C^*-algebra is a normalized cp map.
  (ii) If $V : \mathcal{H} \to \mathcal{K}$ is a bounded linear map between Hilbert spaces, then $\mathcal{F} : \mathfrak{B}(\mathcal{K}) \to \mathfrak{B}(\mathcal{H})$ defined by $\mathcal{F}(a) := V^*aV$ is a cp map. It is normalized if and only if V is an isometry.
 (iii) Let $\mathcal{H}$ be a Hilbert space carrying a continuous unitary representation of a finite dimensional compact Lie group K. Denote the unitaries representing $g \in$ K by $U(g)$, and let $\mathcal{F} : \mathfrak{B}(\mathcal{H}) \to \mathfrak{B}(\mathcal{H})$ be the map

$$\mathcal{F}(a) = \int_K \mathrm{d}g \; U(g)aU(g)^* \tag{3.3}$$

---

[2]Here the tensor product $M_N(\mathbb{C}) \otimes \mathfrak{A}_2$ is algebraic, with no completion required. To obtain the (unique) C^*-norm, one may use the fact that there exists a universal representation $\pi_u : \mathfrak{A}_2 \to \mathfrak{B}(\mathcal{H}_u)$, which is faithful and hence isometric. One may then represent $M_N(\mathbb{C}) \otimes \mathfrak{A}_2$ on $\mathbb{C}^N \otimes \mathcal{H}_u$ and use the operator norm.

where dg is the normalized Haar measure. Then $\mathcal{F}$ is normalized and completely positive, because $1 \otimes \mathcal{F}$ can be written in the same form with $U(g)$ replaced by the representation $1 \otimes U(g)$.

(iv) When $\mathfrak{A}_1$ and $\mathfrak{A}_2$ are v. Neumann algebras, then $\mathcal{F} : \mathfrak{A}_1 \to \mathfrak{A}_1 \otimes \mathfrak{A}_2$ defined by $\mathcal{F}(a) := a \otimes 1_2$ defines a normalized cp map. It corresponds to the restriction of states from $\mathfrak{A}_1 \otimes \mathfrak{A}_2$ to $\mathfrak{A}_1$.

(v) Similarly, given a state ω_2 on $\mathfrak{A}_2$ the map $\mathcal{F} : \mathfrak{A}_1 \otimes \mathfrak{A}_2 \to \mathfrak{A}_1$ defined by $\mathcal{F}(a \otimes b) := a\omega_2(b)$ is a normalized cp map. The corresponding map on states sends ω_1 to $\omega_1 \otimes \omega_2$, which corresponds to attaching an ancillary system $\mathfrak{A}_2$ in the state ω_2. This map is a right-inverse to the restriction map above.

(vi) For each $N > 1$ the map $\mathcal{F} : \mathfrak{B}(\mathbb{C}^N) \to \mathfrak{B}(\mathbb{C}^N)$ defined by $\mathcal{F}(a) = (\text{Tr } a) 1_N - a$ is positive. It is normalized only for $N = 2$. Interestingly, it is *not* completely positive.

(vii) Let ψ_j, $j = 1, \ldots, d$ be operators on a Hilbert space $\mathcal{H}$ satisfying the relations of the Cuntz algebra, see Sect. 2.2.3. Then $\rho : \mathfrak{B}(\mathcal{H}) \to \mathfrak{B}(\mathcal{H})$ given by $\rho(a) = \sum_j \psi_j a \psi_j^*$ is a normalized cp map. It plays a role in the theory of superselection sectors.

A general result due to Stinespring [6] shows that all completely positive maps $\mathcal{F} : \mathfrak{A} \to \mathfrak{B}(\mathcal{H})$ can be written as $\mathcal{F}(a) = V^*\pi(a)V$, where π is a representation of $\mathfrak{A}$ on some Hilbert space $\mathcal{K}$ and $V : \mathcal{H} \to \mathcal{K}$ is bounded. When $\mathcal{F}$ is normalized, one can choose an isometry V. When $\mathfrak{A}$ already acts on $\mathcal{H}$ and $\pi(a) = \oplus_j a$ is a (finite or countable) direct sum representation on $\mathcal{H}^{\oplus N}$ one recovers a formulation in terms of Kraus operators:

$$\mathcal{F}(a) = \sum_j V_j^* a V_j, \qquad \sum_j V_j^* V_j = 1. \tag{3.4}$$

It follows from standard properties of finite type I factors that in this case, all cp maps arise in this way [7], but this is no longer true for general type, in particular type III.

Returning to properties for entanglement measures, we now consider cp maps $\mathcal{F} : \mathfrak{A}_{\hat{A}} \otimes \mathfrak{A}_{\hat{B}} \to \mathfrak{A}_A \otimes \mathfrak{A}_B$. We call such a map **"local"** if it is of the form

$$\mathcal{F}(a \otimes b) = \mathcal{F}_A(a) \otimes \mathcal{F}_B(b) \equiv (\mathcal{F}_A \otimes \mathcal{F}_B)(a \otimes b), \tag{3.5}$$

where the $\mathcal{F}_A$ and $\mathcal{F}_B$ are normal cp maps. More generally, we make the following key definition:

Definition 9 A **"separable operation"** is by definition a family of normal, local cp maps $\mathcal{F}_j$, which are each of tensor product form (3.5), satisfying additionally $\sum_j \mathcal{F}_j(1) = 1$. We think of such an operation as mapping a state ω with probability $p_j := \omega((\mathcal{F}_{A,j} \otimes \mathcal{F}_{B,j})(1))$ to $\frac{1}{p_j}(\mathcal{F}_{A,j} \otimes \mathcal{F}_{B,j})^*\omega$.

It is clear that separable operations map separable states to separable states. In the literature on Quantum Information Theory, it is argued that an arbitrary combination of local operations and "classical communication" (LOCC) between systems A and B is modeled by a separable operation. However, not all separable operations are actually LOCC operations [1]. The notion of an LOCC operation is closer to what actually seems experimentally feasible, and thus conceptually superior. But that notion is also more complicated, and besides, even many LOCC operations may not be experimentally feasible [8]. Moreover, the assumption that A and B can communicate their (classical) measurement results is at any rate inappropriate in a relativistic theory such as QFT, when the regions A and B are spacelike separated, see [9] for a discussion. We thus see that it is perhaps overly restrictive to consider all separable operations, especially in QFT. We will nevertheless do so, since separable operations are rather easy to describe and handle, and we leave a more thorough discussion of this matter to the future.

We can now formulate the idea that on average, no entanglement can be won by performing separable operations:

(e4) (monotonicity under separable operations) Consider a separable operation, described by normal cp maps $\mathcal{F}_j = \mathcal{F}_{A,j} \otimes \mathcal{F}_{B,j}$ with $\sum_j \mathcal{F}_j(1) = 1$. Then

$$\sum_j p_j E\left(\frac{\mathcal{F}_j^* \omega}{p_j}\right) \leq E(\omega), \qquad (3.6)$$

where we sum over all j with $p_j := \omega(\mathcal{F}_j(1)) > 0$.

As examples one can consider some of the cp maps mentioned above. When (e4) holds, E is preserved under the action of local unitaries, i.e. separable operations of the form $\mathcal{F}(a \otimes b) := U_A^* a U_A \otimes U_B^* b U_B$ with unitaries U_A and U_B on $\mathcal{H}_A$ and $\mathcal{H}_B$, respectively, because $\mathcal{F}$ has an inverse which is again a separable operation. Similarly, E is preserved under attaching a local ancillary system, e.g. $\mathcal{F} : (\mathfrak{A}_A \otimes \mathfrak{A}_C) \otimes \mathfrak{A}_B \to \mathfrak{A}_A \otimes \mathfrak{A}_B$ with $\mathcal{F}((a \otimes c) \otimes b) = \omega_C(c)a \otimes b$, because this separable operation has a left inverse. (Note that we need to choose whether the ancillary system is attached to system A or B in order to view $\mathfrak{A}_A \otimes \mathfrak{A}_C \otimes \mathfrak{A}_B$ as a bipartite system and to define separable states and entanglement.) On the other hand, the restriction of states to a subalgebra of the form $\mathfrak{A}_C \otimes \mathfrak{A}_B$ with $\mathfrak{A}_C \subset \mathfrak{A}_A$ may decrease the value of E.

Next we consider what happens when the systems A and B themselves are composed of statistically independent subsystems. In that case, one may wish to ask additionally that (e5) and/or (e6) hold:

(e5) (tensor products) Let $\mathfrak{A}_A = \mathfrak{A}_{A_1} \otimes \mathfrak{A}_{A_2}$ and $\mathfrak{A}_B = \mathfrak{A}_{B_1} \otimes \mathfrak{A}_{B_2}$, and let $\omega_{A_j B_j}$, $j = 1, 2$, be states on $\mathfrak{A}_{A_j} \otimes \mathfrak{A}_{B_j}$. Then

$$E(\omega_{A_1 B_1} \otimes \omega_{A_2 B_2}) \leq E(\omega_{A_1 B_1}) + E(\omega_{A_2 B_2}). \qquad (3.7)$$

(e6) (superadditivity) Let ω_{AB} be a state on $\mathfrak{A}_A \otimes \mathfrak{A}_B$ with $\mathfrak{A}_A = \mathfrak{A}_{A_1} \otimes \mathfrak{A}_{A_2}$ and $\mathfrak{A}_B = \mathfrak{A}_{B_1} \otimes \mathfrak{A}_{B_2}$, and let $\omega_{A_i B_i}$ be its restriction to $\mathfrak{A}_{A_i} \otimes \mathfrak{A}_{B_i}$ (embedded as e.g. $1 \otimes \mathfrak{A}_{A_2} \otimes 1 \otimes \mathfrak{A}_{B_2}$). Then

$$E(\omega_{A_1 B_1}) + E(\omega_{A_2 B_2}) \le E(\omega_{AB}). \tag{3.8}$$

3.3 Bell Correlations As an Entanglement Measure

Historically, the first quantity that was used as a measure of entanglement was the violation of the Bell-inequalities [10, 11]. A convenient formulation is as follows. For commuting subalgebras $\mathfrak{A}_A, \mathfrak{A}_B$ of some v. Neumann algebra $\mathfrak{A}$, and ω a state on $\mathfrak{A}$, we define

$$E_B(\omega) := \sup\{\tfrac{1}{2}\omega(a_1(b_1 + b_2) + a_2(b_1 - b_2))\} \tag{3.9}$$

where the supremum is over all self-adjoint elements a_i, b_i such that

$$a_i \in \mathfrak{A}_A, \quad -1 \le a_i \le 1, \quad b_i \in \mathfrak{A}_B \quad -1 \le b_i \le 1. \tag{3.10}$$

The properties of this quantity in the context of algebraic quantum field theory are discussed e.g. in [12]. It can be demonstrated that $\sqrt{2} \ge E_B(\omega) \ge 1$ [13–15], and that the lower bound is achieved for separable states, so no state with $E_B(\omega) > 1$ can be separable. The equality $E_B(\omega) = 1$ is equivalent to the Clauser-Horne-Shimony-Holt [16] version of Bell's inequalities.

The measure $E_B(\omega)$ of the Bell correlations has several nice properties, including (e0), (e2), (e3) and (e4), but unfortunately it fails (e1). The normalisation $E_B(\omega) \ge 1$ rather than ≥ 0 seems harmless, but the main problem is that there are entangled states ω with $E_B(\omega) = 1$ [17].

There exist many other measures for entanglement and it is impossible to list them all here, but we refer to [18] for an overview (in the type I case). Our main focus will be on the relative entanglement entropy, which will be introduced in the next section. The other measures that we introduce will be useful as convenient tools to derive upper and lower bounds on the relative entanglement entropy.

3.4 Relative Entanglement Entropy

The mother of all notions of entropy is the v. Neumann entropy. It is defined for density matrices ρ on a Hilbert space $\mathcal{H}$ by $H_{vN}(\rho) = -\operatorname{Tr}(\rho \ln \rho)$. The v. Neumann entropy can be viewed as the lack of information about a system to which one has ascribed the state ρ, assuming that the observer has, in principle, access to all operations (observables) in $\mathfrak{B}(\mathcal{H})$. This interpretation is in accord for instance with the facts that $H_{vN}(\rho) \ge 0$ and that a pure state $\rho = |\Psi\rangle\langle\Psi|$ has vanishing v. Neumann entropy.

A related notion is that of the relative entropy. It is defined for two density matrices ρ, ρ' by

$$H(\rho, \rho') = \mathrm{Tr}(\rho \ln \rho - \rho \ln \rho'). \qquad (3.11)$$

The relative entropy can be thought of as the expected amount of information we gain when we update our belief about the state of the system from ρ' to ρ [19]. Like $H_{\mathrm{vN}}(\rho)$, $H(\rho, \rho')$ is non-negative, but can be infinite.

It seems hard to generalize the v. Neumann entropy to algebras of arbitrary type, in particular for type III. But a generalization of the relative entropy to v. Neumann algebras of arbitrary type was found by Araki [20, 21]. It is formulated using modular theory. One assumes to be given two faithful, normal states ω, ω' on a v. Neumann algebra $\mathfrak{A}$ in standard form. We choose the vector representatives in the natural cone $\mathcal{P}^\sharp$, called $|\Omega\rangle, |\Omega'\rangle$ (cf. Proposition 2). Imitating the construction in Sect. 2.1, one defines, following Araki [22], $S_{\omega,\omega'} a |\Omega'\rangle = a^*|\Omega\rangle$, and one considers again the polar decomposition $S_{\omega,\omega'} = J \Delta_{\omega,\omega'}^{1/2}$ (the anti-unitary J is seen to coincide with the corresponding J for the state ω). A related object is the Connes-cocycle (Radon-Nikodym-derivative) defined as $[D\omega : D\omega']_t = \Delta_{\omega,\psi}^{it} \Delta_{\psi,\omega'}^{it} \in \mathfrak{A}$, where ψ is an arbitrary auxiliary faithful state on $\mathfrak{A}'$ (the definition is seen not to depend on it).

Definition 10 The relative entropy is

$$H(\omega, \omega') = \langle \Omega | \ln \Delta_{\omega,\omega'} \, \Omega \rangle = \lim_{t \to 0} \frac{\omega([D\omega : D\omega']_t - 1)}{it}, \qquad (3.12)$$

H is extended to positive functionals that are not necessarily normalized by the formula $H(\lambda\omega, \lambda'\omega') = \lambda H(\omega, \omega') + \lambda \ln(\lambda/\lambda')$, where $\lambda, \lambda' > 0$ and ω, ω' are normalized. If ω' is not normal, then one sets $H(\omega, \omega') = \infty$. When ω or ω' are not faithful (such that $|\Omega\rangle, |\Omega'\rangle$ are not standard), the definition has to be somewhat modified [23].

Key example from Sect. 2.1 continued: For a type I algebra $\mathfrak{A} = \mathfrak{B}(\mathcal{H})$, states ω, ω' correspond to density matrices ρ, ρ'. The relative modular operator $\Delta_{\omega,\omega'}^{1/2}$ corresponds to $\rho^{1/2} \otimes \rho'^{-1/2}$ in the representation of $\mathfrak{A}$ on $\mathcal{H} \otimes \bar{\mathcal{H}}$. In this representation, ω corresponds to the vector state $|\Omega\rangle = \rho^{1/2} \in \mathcal{H} \otimes \bar{\mathcal{H}}$, and the abstract definition of the relative entropy in (3.12) becomes

$$\langle \Omega | \ln \Delta_{\omega,\omega'} \Omega \rangle = \mathrm{Tr}_\mathcal{H} \, \rho^{\frac{1}{2}} \left(\ln \rho \otimes 1 - 1 \otimes \ln \rho' \right) \rho^{\frac{1}{2}} = \mathrm{Tr}_\mathcal{H}(\rho \ln \rho - \rho \ln \rho'), \qquad (3.13)$$

and therefore reproduces that given in (3.11) for density matrices.

Let us now recall the main properties of H (see [23] for a thorough discussion and references).

(h1) (positivity) $H(\omega, \omega') \geq 0$, and $H(\omega, \omega') = 0 \Rightarrow \omega = \omega'$ for states ω, ω'. If ω, ω' are not normal (i.e. their GNS representations are not quasi-equivalent), then $H(\omega, \omega') = \infty$. (The reverse implication is in general false, i.e. the relative entropy can be infinite for normal states).

(h2) (lower semi-continuity) The map $(\omega, \omega') \mapsto H(\omega, \omega')$ is weakly lower semi-continuous on the space of positive functionals on a C^*-algebra.

(h3) (subadditivity) $H(\sum_j \psi_j, \sum_j \varphi_j) \le \sum_j H(\psi_j, \varphi_j)$ for finite sums of normal positive functionals. (Note that this is equivalent to convexity: $H(\sum_j \lambda_j \psi_j, \sum_j \lambda_j \varphi_j) \le \sum_j \lambda_j H(\psi_j, \varphi_j)$ when $\lambda_j \ge 0$ has $\sum_j \lambda_j = 1$.)

(h4) (superadditivity in first argument) $H(\sum_j \omega_j, \omega') \ge \sum_j H(\omega_j, \omega')$ for finite sums of normal positive functionals.

(h5) (monotonicity) If $\phi \le \omega$, $\|\phi\| = \|\omega\|$ and $\phi' \le \omega'$ for normal positive functionals ω', ϕ', ω, ϕ, then $H(\omega, \omega') \le H(\phi, \phi')$.

(h6) ("Uhlmann's monotonicity Theorem" [5, 24]) If $\mathcal{F} : \mathfrak{A}_1 \to \mathfrak{A}_2$ is a normalized cp map between v. Neumann algebras, then $H(\mathcal{F}^*\omega, \mathcal{F}^*\omega') \le H(\omega, \omega')$. Equality holds if $\mathcal{F} = \mathcal{E}$ is a faithful, normal, conditional expectation from $\mathfrak{A}_1$ to a subalgebra $\mathfrak{A}_2$, i.e. $\mathcal{E}(abc) = a\mathcal{E}(b)c$ for $a, c \in \mathfrak{A}_2 \subset \mathfrak{A}_1$, $b \in \mathfrak{A}_1$ and there is a faithful normal state ϕ on $\mathfrak{A}_1$ such that $\phi \circ \mathcal{E} = \phi$ (such a map is always completely positive).

(h7) (tensor product) Let $\mathfrak{A} = \mathfrak{A}_1 \otimes \mathfrak{A}_2$ be the (spatial) tensor product of two v. Neumann algebras, let ω be a normal state on $\mathfrak{A}$ with $\omega_i := \omega|_{\mathfrak{A}_i}$ and let ω'_i be normal states on $\mathfrak{A}_i$. Then $H(\omega, \omega'_1 \otimes \omega'_2) = H(\omega, \omega_1 \otimes \omega_2) + H(\omega_1, \omega'_1) + H(\omega_2, \omega'_2)$.

With the help of the relative entropy we can now define two entanglement measures [3]:

Definition 11 The "relative entanglement entropy" $E_R(\omega)$ of a normal state ω on the tensor product $\mathfrak{A}_A \otimes \mathfrak{A}_B$ of two v. Neumann algebras (in standard form) is given by

$$E_R(\omega) := \inf\{H(\omega, \sigma) \mid \sigma \text{ a separable state}\}. \tag{3.14}$$

The "mutual information" $E_I(\omega)$ is given by

$$E_I(\omega) = H(\omega, \omega_A \otimes \omega_B), \tag{3.15}$$

where $\omega_A := \omega|_{\mathfrak{A}_A}$ and similarly for ω_B.

It immediately follows that $E_R(\omega) \le E_I(\omega)$.

As an example, let us consider a bi-partite system with Hilbert space $\mathcal{H}_A \otimes \mathcal{H}_B$ and observable algebra $\mathfrak{A} = \mathcal{B}(\mathcal{H}_A) \otimes \mathcal{B}(\mathcal{H}_B)$. A pure state ω_{AB} on $\mathfrak{A}$ corresponds to a density matrix $\rho_{AB} = |\Phi\rangle\langle\Phi|$, where $|\Phi\rangle \in \mathcal{H}_A \otimes \mathcal{H}_B$. One calls $\rho_A = \text{Tr}_{\mathcal{H}_B} \rho_{AB}$ the "reduced density matrix", which defines a state ω_A on $\mathcal{B}(\mathcal{H}_A)$ (and similarly for system B). The relative entanglement entropy between A and B in the pure state $\omega \equiv \omega_{AB}$ is then [3]

$$E_R(\omega) = H_{\text{vN}}(\rho_A) \quad [= H_{\text{vN}}(\rho_B)]. \tag{3.16}$$

The mutual information, often used in the case when ω is mixed, i.e. when ρ_{AB} is not a rank 1 projector, is given in our example system by

$$E_I(\omega) = H_{\text{vN}}(\rho_A) + H_{\text{vN}}(\rho_B) - H_{\text{vN}}(\rho_{AB}). \tag{3.17}$$

When $\omega = \mathrm{Tr}(\rho_{AB}\,.)$ is pure, then evidently $E_I(\omega) = 2E_R(\omega)$. If ω is not pure, then $E_I(\omega)$ will be strictly smaller than $2E_R(\omega)$. E_I satisfies (e1) for product states $\omega = \omega_A \otimes \omega_B$, but for general separable states of the form $\omega = \sum_i p_i \omega_{Ai} \otimes \omega_{Bi}$ (with ω_{Ai}, ω_{Bi} states) we can show using (h3), (h5), (h7) only that $E_I(\omega) \leq H_{\mathrm{vN}}(\{p_i\})$. (e3) fails and the status of (e4)-(e6) is the same as for E_R, see below.

In the next sections we investigate these quantities in various algebraic quantum field theories, where we will always take

$$\mathfrak{A}_A \cong \pi(\mathfrak{A}(O_A))'', \quad \mathfrak{A}_B \cong \pi(\mathfrak{A}(O_B))''$$

for two space-like separated open sets A, B in some Cauchy surface with finite distance, where π is a suitable representation (usually the GNS-representation of the state ω considered). We have already noted that these algebras are of type III_1, and so *never have a normal pure state*. Consequently, in our case, we typically expect a strict inequality $E_R(\omega) < E_I(\omega)$. It is also essential in this situation that the regions O_A, O_B must have a finite, positive distance. Otherwise standard states, such as the vacuum, will usually not be normal states on $\mathfrak{A}(O_A) \otimes \mathfrak{A}(O_B)$ (as we have seen in our discussion of the split property, Theorem 2), and hence automatically lead to an infinite relative entanglement entropy by (h1). This phenomenon is indeed encountered in many formal approaches to entanglement entropy in quantum field theory, where one implicitly assumes that the type of the algebra is I.

The properties of the relative entropy directly imply many properties of $E_R(\omega)$, where $\mathfrak{A}_A, \mathfrak{A}_B$ are two v. Neumann algebras in standard form, as before. In particular we have the properties (e0) (manifest), (e3) (from (h3)), and (e5) (from (h7)). Property (e1) holds due to (h1) with the modification that $E_R(\omega) = 0$ implies that ω is a norm limit of separable states (cf. [23] Theorem 5.5). The continuity (e2) was shown for matrix algebras in [25]. The key requirement (e4) does not directly follow from (h4) and (h6), but we can argue as follows adapting somewhat the proof by [3] for type I factors: First, let $\mathfrak{M}$ be any v. Neumann algebra with n normal cp maps $\mathcal{F}_i$ defined on it such that $\sum_i \mathcal{F}_i(1) = 1$. As a technical simplification, we assume that $\mathcal{F}_i(a) = 0, a \in \mathfrak{A}^+$ implies $a = 0$ (this assumption can be removed).

Letting ω, ω' be two faithful normal states, we first show:

Lemma 2 *We have $\sum_i H(\mathcal{F}_i^* \omega, \mathcal{F}_i^* \omega') \leq H(\omega, \omega')$.*

Proof Define $\hat{\mathfrak{M}} = \mathfrak{M} \otimes M_n(\mathbb{C})$. Denoting by $\{|i\rangle\}$ an orthonormal basis of $\mathbb{C}^n$, we define

$$\hat{\mathcal{F}} : \hat{\mathfrak{M}} \to \mathfrak{M}, \quad a \otimes X \mapsto \sum_i \langle i|X|i\rangle \, \mathcal{F}_i(a), \qquad (3.18)$$

which is easily checked to be cp. Using the projections $P_i = 1 \otimes |i\rangle\langle i|, i = 1, \ldots, n$, we also define the cp maps $\mathcal{E}_i : \hat{\mathfrak{M}} \to \hat{\mathfrak{M}}, \mathcal{E}_i(\hat{a}) = P_i \hat{a} P_i$. Using the properties $P_i P_j = \delta_{ij} P_j, \sum_i P_i = 1$, one can show that for any pair of normal states ψ, ψ' on $\hat{\mathfrak{M}}$, one has

$$\sum_i H(\mathcal{E}_i^* \psi, \mathcal{E}_i^* \psi') = H\left(\sum_i \mathcal{E}_i^* \psi, \sum_j \mathcal{E}_j^* \psi'\right). \tag{3.19}$$

This property is obvious for type I factors, and can be proven in the general case as follows. Let $\alpha_i : \mathfrak{M} \to \hat{\mathfrak{M}}$ be the $*$-homomorphisms $\alpha_i(a) = |i\rangle\langle i| \otimes a$. The image is a v. Neumann subalgebra $\hat{\mathfrak{M}}_i$, and $\mathcal{E}_i$ is obviously a faithful conditional expectation onto this subalgebra. Now let $\psi_i = \alpha_i^* \psi$ (and similarly $\psi_i' = \alpha_i^* \psi'$), and let $|\Omega_i\rangle, |\Omega_i'\rangle \in \mathcal{P}^\sharp \subset \mathcal{H}$ be vector representers in a natural cone in a Hilbert space representation of $\mathfrak{M}$ on a Hilbert space $\mathcal{H}$. Letting $\varphi = \sum_i \mathcal{E}_i^* \psi$, we can describe the associated GNS-representation $\hat{\mathcal{H}}_\varphi, |\Omega_\varphi\rangle$ as follows. The Hilbert space is $\hat{\mathcal{H}} = M_n(\mathbb{C}) \otimes \mathcal{H}$ with inner product $\langle X' \otimes \Psi' | X \otimes \Psi\rangle = \mathrm{Tr}(X'^* X)\langle \Psi' | \Psi\rangle_{\mathcal{H}}$. Elements $Y \otimes a \in \hat{\mathfrak{M}}$ act by $Y \otimes a | X \otimes \Psi\rangle = |YX \otimes a\Psi\rangle$. The GNS vector is $|\Omega_\varphi\rangle = \sum_i |i\rangle\langle i| \otimes |\Omega_i\rangle$. Analogous statements hold for $\varphi' = \sum_i \mathcal{E}_i^* \psi'$. If ψ, ψ' are faithful, then both $|\Omega_\varphi\rangle, |\Omega_{\varphi'}\rangle$ are separating, as are $|\Omega_i\rangle, |\Omega_i'\rangle$. The relative modular operator and modular conjugation are found to be

$$\Delta(\varphi, \varphi')^{it}(|k\rangle\langle j| \otimes |\Psi\rangle) = |k\rangle\langle j| \otimes \Delta(\psi_k, \psi_j')^{it}|\Psi\rangle, \quad J(|k\rangle\langle j| \otimes |\Psi\rangle) = |j\rangle\langle k| \otimes J|\Psi\rangle \tag{3.20}$$

where J on the right side of the last equation is the modular conjugation associated with the natural cone $\mathcal{P}^\sharp$. It immediately follows from these formulas and the definition of the relative entropy that

$$H\left(\sum_i \mathcal{E}_i^* \psi, \sum_j \mathcal{E}_j^* \psi'\right) = \sum_i H(\psi_i, \psi_i'). \tag{3.21}$$

However, the terms on the right side can also be written as $\sum_i H(\mathcal{E}_i^* \psi, \mathcal{E}_i^* \psi')$, since $\mathcal{E}_i$ is a faithful conditional expectation onto the subalgebra $\hat{\mathfrak{M}}_i$ and $\mathcal{E}_i^* \psi = \psi|_{\hat{\mathfrak{M}}_i}$, and similarly for ψ', by (h6). The proof of (3.19) is complete.

If we embed $\mathfrak{M}$ into $\hat{\mathfrak{M}}$ as $a \mapsto a \otimes 1_n$, then it follows from the definitions that $\hat{\mathcal{F}}\mathcal{E}_i|_{\mathfrak{M}} = \mathcal{F}_i$. It follows from the definitions that $\sum_i \hat{\mathcal{F}}\mathcal{E}_i = \hat{\mathcal{F}}$. Using these properties, we have, for normal states ω, ω' on $\mathfrak{M}$ (noting that $\hat{\mathcal{F}}^* \omega = \psi, \hat{\mathcal{F}}^* \omega' = \psi'$ are faithful):

$$\begin{aligned}
\sum_i H(\mathcal{F}_i^* \omega, \mathcal{F}_i^* \omega') &= \sum_i H((\hat{\mathcal{F}}\mathcal{E}_i)^* \omega|_{\mathfrak{M}}, (\hat{\mathcal{F}}\mathcal{E}_i)^* \omega'|_{\mathfrak{M}}) \\
&\leq \sum_i H((\hat{\mathcal{F}}\mathcal{E}_i)^* \omega, (\hat{\mathcal{F}}\mathcal{E}_i)^* \omega') \\
&= H(\sum_i (\hat{\mathcal{F}}\mathcal{E}_i)^* \omega, \sum_j (\hat{\mathcal{F}}\mathcal{E}_j)^* \omega') \\
&= H(\hat{\mathcal{F}}^* \omega, \hat{\mathcal{F}}^* \omega') \leq H(\omega, \omega').
\end{aligned} \tag{3.22}$$

To go to the second line, we used (h6) applied to the inclusion of $\mathfrak{M}$ into $\hat{\mathfrak{M}}$. To go to the third line we used (3.19), and in the last step we used (h6) applied to $\hat{\mathcal{F}}$. □
We now have, with $p_i = \omega(\mathcal{F}_i(1))$, $p'_i = \omega'(\mathcal{F}_i(1))$:

$$\sum_i p_i H(\mathcal{F}_i^* \omega / p_i, \mathcal{F}_i^* \omega' / p'_i) = \sum_i H(\mathcal{F}_i^* \omega, \mathcal{F}_i^* \omega') - \sum_i p_i \ln(p_i / p'_i)$$

$$= \sum_i H(\mathcal{F}_i^* \omega, \mathcal{F}_i^* \omega') - H(\mathrm{diag}\{p_i\}, \mathrm{diag}\{p'_i\})$$

$$\leq \sum_i H(\mathcal{F}_i^* \omega, \mathcal{F}_i^* \omega'),$$

$$(3.23)$$

using in the first step the scaling properties of the relative entropy, and using in the last step the property (h1) for the diagonal density matrices $\mathrm{diag}\{p_i\}$, $\mathrm{diag}\{p'_i\}$. We therefore conclude altogether that

$$\sum_i p_i H(\mathcal{F}_i^* \omega / p_i, \mathcal{F}_i^* \omega' / p'_i) \leq H(\omega, \omega'). \qquad (3.24)$$

To show (e4), one now takes $\mathfrak{M} = \mathfrak{A}_A \otimes \mathfrak{A}_B$, and for ω' a separable state σ with the property $E_R(\omega) \geq H(\omega, \sigma) - \epsilon$ for an arbitrary but fixed $\epsilon > 0$. The statement then follows immediately from inequality (3.24) since each $\mathcal{F}_i^* \sigma / \sigma(\mathcal{F}_i(1)) = \sigma_i$ is again a separable state, so that $\sum_i p_i E_R(\mathcal{F}_i^* \omega / p_i) - \epsilon \leq E_R(\omega)$.

3.5 Logarithmic Dominance

We say that a positive linear functional σ on a C^*-algebra $\mathfrak{A}$ is dominated by a positive linear functional σ' when $\sigma' - \sigma$ is positive, i.e. $\sigma' \geq \sigma \geq 0$. Using the properties of positive linear functionals we have

$$\|\sigma' - \sigma\| = \sigma'(1) - \sigma(1) = \|\sigma'\| - \|\sigma\|.$$

In particular, for two states ω and ω' we have $\omega' \geq \omega$ if and only if $\omega' = \omega$.

We will call a positive linear functional σ on $\mathfrak{A}_A \otimes \mathfrak{A}_B$ separable, when $\sigma = r\omega$ with ω a separable state and $r = \sigma(1) \geq 0$.

Using these notions we can now introduce a further entanglement measure, which has been introduced and analyzed in the type I setting in [26], where it was termed "max entropy"[3]:

Definition 12 The "logarithmic dominance" $E_N(\omega)$ of a normal state ω on the tensor product $\mathfrak{A}_A \otimes \mathfrak{A}_B$ of two v. Neumann algebras (in standard form) is given by

---

[3]We thank Marc M. Wilde for pointing out this reference to us.

$$E_N(\omega) := \inf\{\ln(\|\sigma\|) \mid \sigma \geq \omega, \ \sigma \text{ separable}\}. \tag{3.25}$$

If no dominating separable functionals σ exist, we set $E_N(\omega) := \infty$.

E_N satisfies property (e0) in a straightforward way. For a modified version of (e1) we note that $\sigma \geq \omega$ implies that $\|\sigma\| \geq 1$ and hence $E_N(\omega) \geq 0$. When ω is separable we have $E_N(\omega) = 0$. Conversely, when $E_N(\omega) = 0$ there is a sequence σ_n of separable positive linear functionals such that $r_n := \|\sigma_n\| = \|\sigma_n - \omega\| + 1$ converges to 1 as $n \to \infty$. Hence, $r_n^{-1}\sigma_n$ converges in norm to ω and ω is a norm limit of separable states. Finally, when a separable, and hence normal, functional σ dominates ω, then ω is necessarily normal too. Conversely, if ω is not normal, then $E_N(\omega) = \infty$.

The validity of property (e2) is unclear, and property (e3) probably fails, because ln is concave rather than convex.

To prove (e4) we consider a separable operation, described by normal cp maps $\mathcal{F}_j = \mathcal{F}_{A,j} \otimes \mathcal{F}_{B,j}$ with $\sum_j \mathcal{F}_j(1) = 1$. We let ω be any state and we set $p_j := \omega(\mathcal{F}_j(1))$, where we may assume $p_j > 0$. We note that each $\mathcal{F}_j$ maps separable positive functionals σ to separable positive functionals $\mathcal{F}_j^*\sigma$, and when $\sigma \geq \omega$, then $p_j^{-1}\mathcal{F}_j^*\sigma \geq p_j^{-1}\mathcal{F}_j^*\omega$. Furthermore,

$$\sum_j \|\mathcal{F}_j^*\sigma\| = \sum_j \sigma(\mathcal{F}_j^*(1)) = \sigma(1) = \|\sigma\|.$$

Using the concavity of ln we therefore find

$$\sum_j p_j E_N(\mathcal{F}_j^*\omega/p_j) \leq \sum_j p_j \inf\{\ln(\|\mathcal{F}_j^*\sigma/p_j\|) \mid \sigma \geq \omega, \ \sigma \text{ separable}\}$$

$$\leq \inf\left\{\sum_j p_j \ln(\|\mathcal{F}_j^*\sigma\|/p_j) \mid \sigma \geq \omega, \ \sigma \text{ separable}\right\}$$

$$\leq \inf\left\{\ln\left(\sum_j \|\mathcal{F}_j^*\sigma\|\right) \mid \sigma \geq \omega, \ \sigma \text{ separable}\right\}$$

$$= \inf\{\ln(\|\sigma\|) \mid \sigma \geq \omega, \ \sigma \text{ separable}\} = E_N(\omega).$$

To show (e5) it suffices to note that two separable functionals σ_i, $i = 1, 2$, which dominate states ω_i, give rise to a separable functional $\sigma := \sigma_1 \otimes \sigma_2$ with $\sigma \geq \omega_1 \otimes \sigma_2 \geq \omega_1 \otimes \omega_2$ and $\|\sigma\| = \sigma(1) = \|\sigma_1\| \cdot \|\sigma_2\|$. By taking the logarithm and the infimum over the σ_i one then finds (e5).

One can also show a weaker version of the superadditivity (e6). Indeed, if $\sigma \geq \omega$, then the restrictions satisfy $\sigma_i \geq \omega_i$ and $\|\sigma_i\| = \|\sigma\|$. Taking the infimum over σ then yields the modified estimate $E_N(\omega_1) + E_N(\omega_2) \leq 2E_N(\omega)$.

To conclude this section we show the following useful estimate:

Theorem 3 $E_R(\omega) \le E_N(\omega)$.

Proof We choose any separable positive linear functional $\sigma \ge \omega$. (If such σ does not exist, the inequality is trivially true.) We then find

$$E_R(\omega) \le H\left(\omega, \frac{\sigma}{\|\sigma\|}\right) = H(\omega, \sigma) + \ln \|\sigma\|$$
$$\le H(\omega, \omega) + \ln \|\sigma\| = \ln \|\sigma\|,$$

using in the first step the definition of the relative entanglement entropy, using the definition of H in the second step, using $\sigma \ge \omega$ and the monotonicity (h5) of H in the third step, and using $H(\omega, \omega) = 0$ in the last step. Taking the infimum over separable $\sigma \ge \omega$ then yields the desired estimate.   □

3.6   Modular Nuclearity As an Entanglement Measure

Our next entanglement measure makes use of modular operators and is especially convenient in quantum field theories that satisfy a modular nuclearity condition, which is somewhat analogous to the BW-nuclearity condition (a5). We will use it below in applications to integrable models in $1 + 1$ dimensions, as well as for conformal quantum field theories and free field theories in $d + 1$ dimensions.

As usual, we consider a v. Neumann algebra $\mathfrak{A}_A \otimes \mathfrak{A}_B \cong \mathfrak{A}_A \vee \mathfrak{A}_B$ represented in standard form on a Hilbert space $\mathcal{H}$ with fixed natural cone $\mathcal{P}^\sharp$. This cone is associated with some fixed cyclic and separating reference vector, but the definition of our entanglement measure will not depend on it. It is natural to define the algebra associated with the "environment" by $\mathfrak{A}_E = (\mathfrak{A}_A \vee \mathfrak{A}_B)'$. The following *standing assumption* in this section will be made:

Standing assumption All states ω considered are such that their vector representative $|\Omega\rangle \in \mathcal{P}^\sharp$ in the natural cone is cyclic for each of $\mathfrak{A}_A$, $\mathfrak{A}_B$ and $\mathfrak{A}_E$. Due to the Reeh-Schlieder theorem, we are naturally in this situation in the context of quantum field theory on Minkowski space if A, B are open subsets of a Cauchy surface $\mathcal{C}$ and E is the complement of the closure of $A \cup B$ in $\mathcal{C}$. Indeed, if π_0 is the vacuum representation and $\mathfrak{A}_A = \pi_0(\mathfrak{A}(O_A))''$ (and similarly for B and E), the standing assumption holds e.g. for normal states ω with bounded energy if the QFT satisfies (a1)-(a4). The cyclic property of $|\Omega\rangle$ can in this case be interpreted physically as saying that an observer occupying the environment has sufficient control over this state.

Under the standing assumption, $|\Omega\rangle$ is cyclic and separating also for the von Neumann algebras $\mathfrak{A}_B'$ and $\mathfrak{A}_A'$ and separating for $\mathfrak{A}_A$ and $\mathfrak{A}_B$.[4] We let Δ and J be

---

[4] Actually, this is all that is needed in order to define E_M.

the corresponding modular operator and modular conjugation for $\mathfrak{A}_B'$. One can then define the map

$$\Psi^A : \mathfrak{A}_A \to \mathcal{H}, \quad \Psi^A(a) = \mathbf{\Delta}^{\frac{1}{4}} a |\Omega\rangle. \tag{3.26}$$

and likewise for A replaced by B. Here and in the following, we write a for $a \otimes 1$ and b for $1 \otimes b$ to simplify the notation, and likewise we often make the identification of $\mathfrak{A}_A$ with $\mathfrak{A}_A \otimes 1$ and of $\mathfrak{A}_B$ with $1 \otimes \mathfrak{A}_B$.

The **Buchholz partition function** is defined by

$$Z(\omega) = \min(\|\Psi^A\|_1, \|\Psi^B\|_1), \tag{3.27}$$

where we use the 1-nuclear norm and the minimum is taken to get a quantity that is manifestly symmetric under an exchange of A with B.

Definition 13 The modular entanglement measure is defined by

$$E_M(\omega) := \ln(Z(\omega)).$$

When neither Ψ^A nor Ψ^B is nuclear, we set $Z(\omega) = \infty$ and $E_M(\omega) = \infty$.

Remark 1 The distinguished value of $\frac{1}{4}$ is due to the formula $\Delta^\alpha a |\Omega\rangle = J \Delta^{\frac{1}{2}-\alpha} a^* |\Omega\rangle$ for $0 \le \alpha \le \frac{1}{2}$, implying that the 1-norm of the functional $\Delta^\alpha a |\Omega\rangle$ is symmetric about the value $\alpha = \frac{1}{4}$. The results given below also apply to other values of $\alpha \in (0, \frac{1}{2})$. We will formulate our proofs in such a way that this should hopefully be evident, but will not make explicit statements.

Key example from Sect. 2.1 continued: Let $\mathfrak{A}_A = M_n(\mathbb{C}) = \mathfrak{A}_B$ and $|\Omega\rangle = \sum_i \sqrt{p_i} |i\rangle\langle i| \in \mathcal{H} = \mathbb{C}^n \otimes \bar{\mathbb{C}}^n$, which is a pure state on $\mathfrak{A}_A \otimes \mathfrak{A}_B$ with corresponding functional $\omega = \langle \Omega| \, . \, |\Omega\rangle$. $|\Omega\rangle$ is therefore not cyclic for $\mathfrak{A}_E$, but it is still cyclic and separating for $\mathfrak{A}_B', \mathfrak{A}_A'$ if all $p_i > 0$. Thus, E_M can still be defined in this case. Going through the definitions, one finds $\Psi^A(a) = \rho^{1/4} a \rho^{1/4}$ and $E_M(\omega) = 2 \ln \mathrm{Tr} \, \rho^{1/4}$, where $\rho = diag(\{p_i\})$, with the same expressions holding also for $B \leftrightarrow A$. For the maximally entangled state ω_n^+ defined by $p_i = 1/n$, we thereby get $E_M(\omega_n^+) = \frac{3}{2} \ln n$.

Our main use of E_M is the following theorem:

Theorem 4 $E_N(\omega) \le E_M(\omega)$.

The proof is based on two lemmas[5]:

Lemma 3 *If $\nu := \|\Psi^A\|_1 < \infty$ and $\epsilon > 0$, there are sequences of (not necessarily positive) normal linear functionals ϕ_j on $\mathfrak{A}_A$ and ψ_j on $\mathfrak{A}_B$ such that*

---

[5] After this preprint appeared, it was pointed out to the authors that a similar proof of Lemma 4 also appears in the unpublished manuscript [27].

$$\omega(ab) = \sum_j \phi_j(a)\psi_j(b) , \qquad a \in \mathfrak{A}_A , \ b \in \mathfrak{A}_B \tag{3.28}$$

and $\sum_j \|\phi_j\| \cdot \|\psi_j\| < \nu + \epsilon.$

Proof Recall that $J|\Omega\rangle = |\Omega\rangle$, and note that $J\Delta^{-\frac{1}{2}}$ is the Tomita operator for $\mathfrak{A}_B$, i.e.

$$\Delta^{-\frac{1}{2}}b^*|\Omega\rangle = J(J\Delta^{-\frac{1}{2}})b^*|\Omega\rangle = Jb|\Omega\rangle = JbJ|\Omega\rangle.$$

Using the commutativity of $\mathfrak{A}_A$ and $\mathfrak{A}_B$ we then note that

$$\begin{aligned}
\omega(ab) &= \langle\Omega|ab\Omega\rangle = \langle(\Delta^{\frac{1}{4}} + \Delta^{-\frac{1}{4}})^{-1}(1 + \Delta^{-\frac{1}{2}})b^*\Omega|\Delta^{\frac{1}{4}}a\Omega\rangle \\
&= \langle(\Delta^{\frac{1}{4}} + \Delta^{-\frac{1}{4}})^{-1}(b^* + JbJ)\Omega|\Psi^A(a)\rangle.
\end{aligned}$$

If $\nu < \infty$ and $\epsilon > 0$, there are sequences of normal functionals ϕ_j on $\mathfrak{A}_A$ and vectors $|\chi_j\rangle \in \mathcal{H}$ such that

$$\Psi^A(a) = \Delta^{\frac{1}{4}}a|\Omega\rangle = \sum_j |\chi_j\rangle\phi_j(a)$$

for all $a \in \mathfrak{A}_A$, and $\sum_j \|\phi_j\| \cdot \|\chi_j\| < \nu + \epsilon$. Define the normal functionals ψ_j on $\mathfrak{A}_B$ by

$$\psi_j(b) := \langle(\Delta^{\frac{1}{4}} + \Delta^{-\frac{1}{4}})^{-1}(b^* + JbJ)\Omega|\chi_j\rangle$$

and note that $\|\psi_j\| \leq \|\chi_j\|$, because $\|(\Delta^{\frac{1}{4}} + \Delta^{-\frac{1}{4}})^{-1}\| \leq \frac{1}{2}$ by the spectral calculus.

Putting both paragraphs together we find the conclusion: $\omega(ab) = \sum_j \phi_j(a)\psi_j(b)$ with $\sum_j \|\phi_j\| \cdot \|\psi_j\| < \nu + \epsilon$. $\qquad\qquad\square$

Lemma 4 *If there are sequences of (not necessarily positive) normal linear functionals ϕ_j on $\mathfrak{A}_A$ and ψ_j on $\mathfrak{A}_B$ such that*

$$\omega(ab) = \sum_j \phi_j(a)\psi_j(b) , \qquad a \in \mathfrak{A}_A , \ b \in \mathfrak{A}_B \tag{3.29}$$

and $\mu := \sum_j \|\phi_j\| \cdot \|\psi_j\| < \infty$, then there is a separable positive linear functional σ such that $\sigma \geq \omega$ and $\|\sigma\| = \mu$.

Proof By Theorem 7.3.2 in [28] there are partial isometries $U_j \in \mathfrak{A}_A$ such that $\phi_j(U_j \ . \ ) \geq 0$ on $\mathfrak{A}_A$ and $\phi_j(U_jU_j^* \ . \ ) = \phi_j$. It follows in particular that $\phi_j(U_j) = \|\phi_j(U_j \ . \ )\| = \|\phi_j\|$ and

$$\bar{\phi}_j(a) = \overline{\phi_j(U_jU_j^*a^*)} = \phi_j(U_j(U_j^*a^*)^*) = \phi_j(U_jaU_j)$$

for all $a \in \mathfrak{A}_A$, where we used the fact that $\phi_j(U_j \cdot)$ is hermitean. (Here $\bar{\psi}(a) \equiv \overline{\psi(a^*)}$.) Similarly, there are partial isometries $V_j \in \mathfrak{A}_B$ such that $\psi_j(V_j \cdot) \geq 0$ and $\psi_j(V_j V_j^* \cdot) = \psi_j$.

Note that the positive linear functional $\rho_j := \phi_j(U_j \cdot) \otimes \psi_j(V_j \cdot)$ is separable. Writing $W_j := U_j \otimes V_j$ we then define

$$\sigma_j := \frac{1}{2}\rho_j + \frac{1}{2}\rho_j(W^* \cdot W) ,$$

which is also separable, because W is a simple tensor product. Furthermore,

$$\|\sigma_j\| = \sigma_j(1) = \rho_j(1) = \|\phi_j\| \cdot \|\psi_j\|$$

and

$$0 \leq \frac{1}{2}\rho_j((1 - W^*) \cdot (1 - W)) = \sigma_j - \frac{1}{2}(\phi_j \otimes \psi_j + \bar{\phi}_j \otimes \bar{\psi}_j) .$$

We conclude that $\sigma := \sum_j \sigma_j$ is a separable positive linear functional with $\|\sigma\| = \sigma(1) = \sum_j \|\sigma_j\| = \mu$ and

$$\sigma \geq \frac{1}{2}\sum_j (\phi_j \otimes \psi_j + \bar{\phi}_j \otimes \bar{\psi}_j) = \frac{1}{2}(\omega + \overline{\omega}) = \omega .$$

$\square$

Proof of Theorem 4: Combining the two lemmas with $\mu = \nu + \epsilon$ we find $E_N(\omega) \leq \ln(\|\sigma\|) = \ln(\mu) = \ln\left(\|\Psi^A\|_1 + \epsilon\right)$ for all $\epsilon > 0$, and hence $E_N(\omega) \leq \ln\left(\|\Psi^A\|_1\right)$. Interchanging the roles of A and B we also find $E_N(\omega) \leq \ln\left(\|\Psi^B\|_1\right)$ and hence $E_N(\omega) \leq \ln(Z(\omega)) = E_M(\omega)$. $\square$

Although we will use E_M only via its relationship to E_N given in Theorem 4, it is perhaps of interest to investigate E_M in its own right. Here we look at properties (e1)-(e6). (e0) is clearly satisfied by construction. (e1) holds in the restricted sense that $E_M(\omega_p) = 0$ when ω is a product state $\omega_A \otimes \omega_B$, which follows immediately from the fact that $\Psi^A(a) = |\Omega\rangle\omega(a)$ in this case. For more general separable states (e1) probably fails. (e2) is unclear to us, but we have the following result, which also implies some sort of continuity different from (e2) for E_M:

Proposition 3 *Let ω_i, $i = 1, 2$ be two faithful normal states on $\mathfrak{A}_A \vee \mathfrak{A}_B \cong \mathfrak{A}_A \otimes \mathfrak{A}_B$, with GNS representers that are separating for $\mathfrak{A}'_A$ and $\mathfrak{A}'_B$ and such that, for some $\lambda > 0$, $\omega_2 \leq \lambda\omega_1$. Then $E_M(\omega_2) - E_M(\omega_1) \leq \frac{1}{2}\ln\lambda$.*

Proof Let $|\Omega_i\rangle$, $i = 1, 2$ be the GNS vector representatives of the states ω_i in $\mathcal{P}^\sharp$, see Proposition 2, and let S_i be the Tomita operators for the algebra $\mathfrak{A}'_B$ associated with $|\Omega_i\rangle$, $i = 1, 2$, with polar decompositions $S_i = J_i\Delta_i^{1/2}$. Note that $\lambda \geq 1$ by the normalisation of states and that $\omega_2 \leq \mu\omega_1$ for any $\mu > \lambda$. We may therefore write

$\omega_1 = \frac{1}{\mu}\omega_2 + \frac{\mu-1}{\mu}\omega_3$, where ω_3 is a normal state. Because $\omega_3 \geq \frac{\mu-\lambda}{\lambda(\mu-1)}\omega_2$ we see that ω_3 is also faithful.

We now need the following important result, which is proved in [20] (Sect. 4). Alternatively, it follows from the "quadratic interpolations" of [24] (Proposition 8).

Lemma 5 *Let $\alpha \in (0, \frac{1}{2})$, and let $\mathfrak{M}$ be a v. Neumann algebra acting on $\mathcal{H}$ with fixed natural cone $\mathcal{P}^{\sharp}$. Then the functional $\omega \mapsto \|\Delta_{\omega}^{\alpha}a\Omega\|^2$ on normal faithful states ω on $\mathfrak{M}$ (with vector representatives $|\Omega\rangle \in \mathcal{P}^{\sharp}$) is concave.*

Applying Lemma 5 we find that for every $a \in \mathfrak{A}_A$,

$$\frac{1}{\mu}\|\Delta_2^{\frac{1}{4}}a\Omega_2\|^2 \leq \|\Delta_1^{\frac{1}{4}}a\Omega_1\|^2,$$

i.e. $\|\Psi_2^A(a)\| \leq \sqrt{\mu}\|\Psi_1^A(a)\|$. This immediately implies that there is an operator T such that $T\Psi_1^A(a) = \Psi_2^A(a)$ and $\|T\| \leq \sqrt{\mu}$. We can therefore estimate

$$\|\Psi_2^A\|_1 = \|T\Psi_1^A\|_1 \leq \|T\|\|\Psi_1^A\|_1 \leq \sqrt{\mu}\|\Psi_1^A\|.$$

A similar estimate holds when we swap the roles of A and B, so we have

$$E_M(\omega_2) - E_M(\omega_1) \leq \ln(\sqrt{\mu}) = \frac{1}{2}\ln(\mu).$$

Taking $\mu \to \lambda^+$ yields the result.   □

Remark 2 When $\mathfrak{M} = M_N$ is the algebra of N dimensional matrices, Lemma 5 states that the map $\rho \mapsto \mathrm{Tr}(\rho^{1-2\alpha}a^*\rho^{2\alpha}a)$ from density matrices ρ to the reals is concave for any $a \in M_N$. This is a special case of the Wigner-Yanase-Dyson-Lieb concavity theorem.

In a similar way we find the converse to (e3), i.e. the concavity of E_M.

Proposition 4 *Let ω_i be normal states on $\mathfrak{A}_A \otimes \mathfrak{A}_B$, and let $\omega = \sum_i \lambda_i\omega_i$ with $\lambda_i > 0$, $\sum_i \lambda_i = 1$. Then $\sum_j \lambda_j E_M(\omega_j) \leq E_M(\omega)$.*

Proof Let $|\Omega\rangle$ be the vector representative for the normal state ω in the chosen natural cone $\mathcal{P}^{\sharp}$ of $\mathfrak{A}_A \otimes \mathfrak{A}_B$. Likewise we have vector representatives $|\Omega_i\rangle$ of ω_i in this cone, and by our standing assumption all these vectors are cyclic for $\mathfrak{A}'_B$. Their modular operators are denoted by Δ_i. Applying Lemma 5 to $\mathfrak{M} = \mathfrak{A}'_B$ with $\alpha = 1/4$ and using the concavity of $x \mapsto \sqrt{x}$ we find

$$\sum_{i=1}^{n} \lambda_i\|\Delta_i^{\frac{1}{4}}a\Omega_i\| \leq \|\Delta^{\frac{1}{4}}a\Omega\| \tag{3.30}$$

for all $a \in \mathfrak{A}_A$. In terms of the maps $\Psi_i^A(a) = \Delta_i^{1/4}a|\Omega_i\rangle$, and $\Psi^A(a) = \Delta^{1/4}a|\Omega\rangle$, this evidently means $\sum_i \lambda_i\|\Psi_i(a)\| \leq \|\Psi(a)\|$. Now let $\Phi^A : \mathfrak{A}_A \to \mathcal{Y} = \oplus^n\mathcal{H}$

be defined by $\Phi^A(a) = (\lambda_1 \Psi_1(a), \ldots, \lambda_n \Psi_n(a))$, and equip $\mathcal{Y}$ with the Banach space norm $\|Y\|_{\mathcal{Y}} = \sum_i \|y_i\|$ for $Y = (y_1, \ldots, y_n) \in \mathcal{Y}$. Then we obviously have $\|\Phi^A(a)\|_{\mathcal{Y}} \leq \|\Psi^A(a)\|$ for all $a \in \mathfrak{A}_A$. It follows that the relation $\{(\Psi^A(a), \Phi^A(a)) \mid a \in \mathfrak{A}_A\} \subset \mathcal{H} \times \mathcal{Y}$ is the graph of a closed linear operator $T : \mathcal{H} \to \mathcal{Y}$ with the property that $\|T\| \leq 1$ and $T \circ \Psi^A = \Phi^A$. Consequently, by the properties of the 1-norm, $\|\Phi^A\|_1 \leq \|\Psi^A\|_1$.

It is elementary to show that $\|\Phi^A\|_1 \geq \sum_i \lambda_i \|\Psi_i^A\|_1$: Suppose $\Phi^A(a) = \sum_\alpha Y_\alpha \varphi_\alpha(a)$ for all $a \in \mathfrak{A}_A$, with $Y_\alpha = (y_{1,\alpha}, \ldots, y_{n,\alpha}) \in \mathcal{Y}$ and normal functionals φ_α chosen such that $\|\Phi^A\|_1 + \epsilon \geq \sum_\alpha \|Y_\alpha\|_{\mathcal{Y}} \|\varphi_\alpha\|$. Obviously, $\lambda_i \Psi_i^\alpha(a) = \sum_\alpha y_{i,\alpha} \varphi_\alpha(a)$, so $\lambda_i \|\Psi_i^A\|_1 \leq \sum_\alpha \|y_{i,\alpha}\| \|\varphi_\alpha\|$ and taking the sum over i it follows that $\|\Phi^A\|_1 + \epsilon \geq \sum_i \lambda_i \|\Psi_i^A\|_1$, and from this the claim follows since ϵ can be made arbitrarily small.

Thus, we know $\|\Psi^A\|_1 \geq \sum_i \lambda_i \|\Psi_i^A\|_1$, and we get the analogous statement for A replaced by B. Taking the ln using its concavity, and taking the minimum over A, B yields the statement. $\qquad\square$

Let us next discuss (e4). Although the next lemma is a special case of the following proposition, we include it here, because the proof is independent.

Lemma 6 *Let* $\mathfrak{A}_{A_1} \subset \mathfrak{A}_{A_2}, \mathfrak{A}_{B_1} \subset \mathfrak{A}_{B_2}$ *let* ω *be a normal state on* $\mathfrak{A}_{A_2} \otimes \mathfrak{A}_{B_2}$ *satisfying our standing assumption. Then* $E_M(\omega \restriction_{\mathfrak{A}_{A_1} \otimes \mathfrak{A}_{B_1}}) \leq E_M(\omega)$.

Proof We let S_i be the Tomita operators for $\mathfrak{A}'_{B_i}$ with polar decompositions $S_i = J_i \Delta_i^{1/2}$. Note that, since $\mathfrak{A}'_{B_2} \subset \mathfrak{A}'_{B_1}$, $\mathrm{dom}(S_2) \subset \mathrm{dom}(S_1)$. Let $\lambda > 0$. The set $\mathrm{dom}(S_1)$ is a Hilbert space called $\mathcal{H}_1$ with respect to the inner product $\langle \Phi | \Psi \rangle_\lambda = \langle \Phi | \Psi \rangle + \lambda^{-1} \langle S_1 \Psi | S_1 \Phi \rangle$, where $\lambda > 0$. Letting $I : \mathcal{H}_1 \to \mathrm{dom}(S_1)$ be the identification map, one shows that $I^{-1} \mathrm{dom}(S_2)$ is a closed subspace $\mathcal{H}_2 \subset \mathcal{H}_1$ with associated orthogonal projection P_2. [29] shows that $I P_i I^* = (1 + \lambda^{-1} \Delta_i)^{-1}$ (with $P_1 = 1$) and that $I^* = I^{-1}(1 + \lambda^{-1} \Delta_1)^{-1}$. It follows for all $b \in \mathfrak{A}'_{B_2}$ that

$$\langle \Omega | b^*(\lambda + \Delta_1)^{-1} b\Omega \rangle - \langle \Omega | b^*(\lambda + \Delta_2)^{-1} b\Omega \rangle = \lambda \|(P_1 - P_2) I^{-1}(\lambda + \Delta_1)^{-1} b\Omega\|^2. \tag{3.31}$$

A standard trick in such a situation is to use the identity $(1 \geq \alpha > 0, t > 0)$

$$t^\alpha = \frac{\sin \pi \alpha}{\pi} \int_0^\infty d\lambda [\lambda^{\alpha-1} - \lambda^\alpha(\lambda + t)^{-1}]. \tag{3.32}$$

Then, if we multiply (3.31) by λ^α, and integrate against λ, we find via the spectral calculus

$$\|\Delta_1^{\frac{\alpha}{2}} b\Omega\|^2 - \|\Delta_2^{\frac{\alpha}{2}} b\Omega\|^2$$
$$= -\frac{\sin \pi \alpha}{\pi} \int_0^\infty d\lambda \, \lambda^{1+\alpha} \|(1 - P_2) I^{-1}(\lambda + \Delta_1)^{-1} b\Omega\|^2 \leq 0. \tag{3.33}$$

For $\alpha = 1/2$, we get[6] $\|\Delta_1^{\frac{1}{4}} b\Omega\| \leq \|\Delta_2^{\frac{1}{4}} b\Omega\|$ for all $b \in \mathfrak{A}'_{B_2}$. This entails the existence of an operator T with $\|T\| \leq 1$ such that $\Delta_1^{\frac{1}{4}} b|\Omega\rangle = T\Delta_2^{\frac{1}{4}} b|\Omega\rangle$ for all $b \in \mathfrak{A}'_{B_2}$.

Since $\mathfrak{A}_{A_1} \subset \mathfrak{A}_{A_2} \subset \mathfrak{A}'_{B_2}$ this relation holds for $a \in \mathfrak{A}_{A_1}$ and we get from the definition of the map Ψ^{A_i} given above in Eq. (3.26) that $\|\Psi^{A_1}\|_1 \leq \|T\| \|\Psi^{A_2}|_{\mathfrak{A}_{A_1}}\|_1 \leq \|\Psi^{A_2}\|_1$. Since the same relation also holds with A replaced by B, we get $E_M(\mathcal{F}^*\omega) \leq E_M(\omega)$ for the embedding $\mathcal{F} : \mathfrak{A}_{A_1} \otimes \mathfrak{A}_{B_1} \to \mathfrak{A}_{A_2} \otimes \mathfrak{A}_{B_2}$, which is the claimed special case of (e4). $\qquad \square$

A more general, but still special, case of (e4) arises when $\mathcal{F}$ is a unit-preserving *-homomorphism ρ of $\mathfrak{A}_A \otimes \mathfrak{A}_B \cong \mathfrak{A}_A \vee \mathfrak{A}_B$ such that $\rho(\mathfrak{A}_A) \subset \mathfrak{A}_A$, and likewise for A replaced by B. Such "localized endomorphisms" arise naturally in the context of the DHR-theory of superselection sectors (charged states) in QFT, see Sect. 4.7. More generally we may consider a homomorphism $\rho : \mathfrak{A}_{A_1} \vee \mathfrak{A}_{B_1} \to \mathfrak{A}_{A_2} \vee \mathfrak{A}_{B_2}$, or finite families thereof.

Proposition 5 E_M satisfies $\sum_i p_i E_M(\rho_i^*\omega) \leq E_M(\omega)$ (here $\sum_i p_i = 1$, $p_i \geq 0$) for localized homomorphisms $\rho_i : \mathfrak{A}_{A_1} \vee \mathfrak{A}_{B_1} \to \mathfrak{A}_{A_2} \vee \mathfrak{A}_{B_2}$ such that each $\omega_i = \rho_i^*\omega$ satisfies our standing assumption for $\mathfrak{A}_{A_1} \vee \mathfrak{A}_{B_1}$, and ω that for $\mathfrak{A}_{A_2} \vee \mathfrak{A}_{B_2}$.

Proof Consider first a single localized endomorphism ρ. Let $|\Omega_\omega\rangle$, $|\Omega_{\rho^*\omega}\rangle$ be the vector representatives of ω, $\rho^*\omega$ in $\mathcal{P}^\sharp$. It follows from the properties of ρ that the linear operator V defined by $Vx|\Omega_{\rho^*\omega}\rangle = \rho(x)|\Omega_\omega\rangle$, $x \in \mathfrak{A}_{A_1} \vee \mathfrak{A}_{B_1}$ is an isometry, $V^*V = 1$. Next, let S_ω, $S_{\rho^*\omega}$ be the Tomita operators for $|\Omega_\omega\rangle$, $|\Omega_{\rho^*\omega}\rangle$ for the v. Neumann algebras $\mathfrak{A}_{B_2}$, $\mathfrak{A}_{B_1}$. The trivial calculation

$$S_\omega Vb|\Omega_{\rho^*\omega}\rangle = \rho(b)^*|\Omega_\omega\rangle = \rho(b^*)|\Omega_\omega\rangle = Vb^*|\Omega_{\rho^*\omega}\rangle = VS_{\rho^*\omega}b|\Omega_{\rho^*\omega}\rangle \quad (3.34)$$

for all $b \in \mathfrak{A}_{B_1}$ establishes the operator equality $S_\omega V = VS_{\rho^*\omega}$ on the domain of $S_{\rho^*\omega}$. By taking adjoints we find on the form domain of $\Delta_{\rho^*\omega}$

$$V^*\Delta_\omega V = \Delta_{\rho^*\omega}, \quad (3.35)$$

where Δ_ω is the modular operator for $\mathfrak{A}_{B_2}$ and the state ω and similarly for $\rho^*\omega$. By the Heinz-Löwner theorem [31], the function $\mathbb{R}_+ \ni x \mapsto x^\alpha$ is operator monotone[7] for $0 < \alpha \leq 1$, so we get $(V^*\Delta_\omega V)^\alpha \leq \Delta_{\rho^*\omega}^\alpha$. Now we need the following result (see e.g. Theorem 2.6 and 4.19 of [32]):

Lemma 7 Let $f : \mathbb{R} \to \mathbb{R}$ be an operator monotone function, V an operator such that $\|V\| \leq 1$, A a positive operator on $\mathcal{H}$. Then $V^*f(A)V \leq f(V^*AV)$ on the form domain of $f(V^*AV)$.

---

[6]For an alternative argument, see Lemma 2.9 of [30], which uses a generalization of the Heinz-Löwner theorem [31] to unbounded operators.

[7]A function $f : \mathbb{R} \to \mathbb{R}$ is called operator monotone if $f(A) \leq f(B)$ whenever two self-adjoint operators A, B on a Hilbert space $\mathcal{H}$ satisfy $A \leq B$ on the form domain of B. If $A = B$ on the form domain of B we obtain $f(A) \leq f(B)$. Notice especially the asymmetry in the assumption on the form domain.

If we apply this lemma to V and $A = \Delta_\omega$, we get $V^* \Delta_\omega^\alpha V \leq (V^* \Delta_\omega V)^\alpha = \Delta_{\rho^* \omega}^\alpha$. This is the same as saying that

$$\Delta_{\rho^* \omega}^{-\alpha/2} V^* \Delta_\omega^{\alpha/2} (\Delta_{\rho^* \omega}^{-\alpha/2} V^* \Delta_\omega^{\alpha/2})^* \leq 1. \tag{3.36}$$

We use this with $\alpha = 1/2$ and take the norm, which gives

$$\| \Delta_{\rho^* \omega}^{-\frac{1}{4}} V^* \Delta_\omega^{\frac{1}{4}} \| \leq 1. \tag{3.37}$$

The modular operator for $\mathfrak{A}'_{B_2}$ required in the definition of Ψ_ω^A (Eq. (3.26)) is $J_\omega \Delta_\omega J_\omega = \Delta_\omega^{-1}$, and similarly for $\rho^* \omega$. It follows using the definition (3.26) that

$$\Psi_{\rho^* \omega}^A(a) = \Delta_{\rho^* \omega}^{-\frac{1}{4}} a |\Omega_{\rho^* \omega}\rangle = \Delta_{\rho^* \omega}^{-\frac{1}{4}} V^* V a |\Omega_{\rho^* \omega}\rangle = T \Delta_\omega^{-\frac{1}{4}} \rho(a) |\Omega_\omega\rangle = T \circ \Psi_\omega^A \circ \rho(a) \tag{3.38}$$

for all $a \in \mathfrak{A}_{A_1}$, where $T = \Delta_{\rho^* \omega}^{-\frac{1}{4}} V^* \Delta_\omega^{\frac{1}{4}}$. The properties of the 1-norm then give $\|\Psi_{\rho^* \omega}^A\|_1 \leq \|\rho\| \|T\| \|\Psi_\omega^A\|_1 \leq \|\Psi_\omega^A\|_1$ and the same for A replaced by B. It follows that $Z(\rho^* \omega) \leq Z(\omega)$ and hence that $E_M(\rho^* \omega) \leq E_M(\omega)$.

Consider next a finite family of localized endomorphisms ρ_i. In this case, the result immediately follows from the previous result and the concavity of ln as (with $\omega_i = \rho_i^* \omega$ and using $Z(\omega_i) \leq Z(\omega)$ for the Buchholz partition function): $\sum_i p_i E_M(\rho_i^* \omega) = \sum_i p_i \ln Z(\omega_i) \leq \ln \sum_i p_i Z(\omega_i) \leq \ln Z(\omega) = E_M(\omega)$. $\qquad\qquad \square$

This concludes our discussion of (e4). Whether the general case of (e4) holds for families of separable operations is unknown to us. Perhaps one could say that at any rate, the properties expressed by Proposition 5 and Lemma 6 are the more natural ones in the context of QFTs. Property (e5) is satisfied since the modular operator behaves functorially under tensor products. Property (c6) is not obvious to us.

3.7 Distillable Entanglement

The last measure that we will discuss is closely related to "entanglement distillation" [9, 33] and is maybe the most natural of all entanglement measures. This measure is formulated in terms of **maximally entangled states**, which are defined for finite dimensional type I factors of the form $\mathfrak{A}_A \otimes \mathfrak{A}_B$, with $\mathfrak{A}_A = M_n(\mathbb{C}) = \mathfrak{A}_B$. Their density matrix is

$$P_n^+ = |\Phi^+\rangle\langle\Phi^+|, \quad |\Phi^+\rangle = \frac{1}{\sqrt{n}} \sum_{i=1}^n |i\rangle \otimes |i\rangle, \tag{3.39}$$

where $\{|i\rangle\}$ is a chosen orthonormal basis of $\mathbb{C}^n$. The corresponding linear functional is denoted by $\omega_n^+ = \mathrm{Tr}(P_n^+ \,.\,)$ in the following. In view of (e4) it is justified to think of these states as maximally entangled because one can show [18] that any other state ω

(pure or mixed) on $\mathfrak{A}_A \otimes \mathfrak{A}_B$ can be obtained as $\omega = \mathcal{F}^* \omega_n^+$ by a separable operation, i.e. a normalized cp map $\mathcal{F}$ on $\mathfrak{A}_A \otimes \mathfrak{A}_B$ that is a convex linear combination of local cp maps of the form (3.5).

The relative entanglement entropy of the maximally entangled state is for instance given by $E_R(\omega_n^+) = \ln n$, as one can see using that $E_R(\omega) = H_{\text{vN}}(\omega_A) = H_{\text{vN}}(\omega_B)$ for all pure states ω. Since $E_R(\omega) \leq E_N(\omega)$ in general, we also have $E_N(\omega_n^+) \geq \ln n$, whereas from Lemma 4, we easily get $E_N(\omega_n^+) \leq \ln n$, implying equality, $E_N(\omega_n^+) = \ln n$. Furthermore, by construction $E_I(\omega_n^+) = 2\ln n$, and $E_B(\omega_n^+) = \sqrt{2}$, since ω_n^+ can be mapped to a Bell-state by a local operation.

Now let $\mathfrak{A}_A, \mathfrak{A}_B$ be general v. Neumann algebras and ω a normal state on $\mathfrak{A} = \mathfrak{A}_A \otimes \mathfrak{A}_B$. The idea of distillation is to take a large number N of copies of this bipartite system $\omega^{\otimes N}$ and "distill" from this ensemble a maximally entangled state ω_n^+ – for as large an $n = n_N$ depending on N as we can – by separable operations. More precisely, we consider sequences $\{n_N\}$ of natural numbers and sequences $\{\mathcal{F}_N\}$ of separable operations, i.e. normalized cp maps $\mathcal{F}_N : M_{n_N}(\mathbb{C}) \otimes M_{n_N}(\mathbb{C}) \to \mathfrak{A}_A^{\otimes N} \otimes \mathfrak{A}_B^{\otimes N} = \mathfrak{A}^{\otimes N}$ that are each convex linear combinations of cp maps of the form (3.5) and have $\mathcal{F}_N(1) = 1$, such that

$$\|\mathcal{F}_N^* \omega^{\otimes N} - \omega_{n_N}^+\| \to 0 \quad \text{as} N \to \infty. \tag{3.40}$$

If such sequences exist, then we call ω **"distillable"** and the sequence $\{\mathcal{F}_N\}$ a "distillation protocol". (This and the following notions clearly do not depend on the choice of basis made above since basis rotations can be implemented by separable operations).

The notion of distillable entropy captures the efficiency of this process. Since $\ln n_N$ is the relative entanglement entropy of the reference state $\omega_{n_N}^+$, the distillation process would be considered as rather inefficient if $\ln n_N \ll N$ asymptotically, while we would consider the distillation process to possess a finite rate if $\ln n_N \propto N$ asymptotically, and the rate itself would be the proportionality constant. The entanglement achieved by the distillation processes $\{\mathcal{F}_N\}$ is defined to be this rate, i.e. we set[8]

$$E_{\{\mathcal{F}_N\}}(\omega) = \limsup_{N \to \infty} \frac{\ln n_N}{N}. \tag{3.41}$$

The distillable entropy is the optimum rate achievable by any such process, i.e. one defines:

Definition 14 The distillable entropy is defined by $E_D(\omega) = \sup_{\{\mathcal{F}_N\}} E_{\{\mathcal{F}_N\}}(\omega)$.

Some general properties of $E_D(\omega)$ are immediately clear from the definition. For instance, we clearly have (e0) and also $E_D(\omega) \geq 0$. (e2) is fairly obvious from the definition, too. We prove (e4) as a lemma:

---

[8]We may always pass to a new protocol whose rate is arbitrarily close, so that the lim sup is actually a lim.

Lemma 8 $E_D(\omega)$ *satisfies (e4), i.e. for any separable operation* $\{\mathcal{E}_i\}$ *with* $\sum_i \mathcal{E}_i(1) = 1$, $\omega(\mathcal{E}_i(1)) = p_i > 0$, *we have* $\sum_i p_i E_D(p_i^{-1}\mathcal{E}_i^*\omega) \leq E_D(\omega)$.

Proof For local operations, i.e. normalized cp maps $\mathcal{E}$ of the form (3.5), this is immediate. For general separable operations, let $\{\mathcal{F}_{i,N}\}$ be near optimal distillation protocols for $\omega_i = p_i^{-1}\mathcal{E}_i^*\omega$ with rates r_i within ϵ of $E_D(\omega_i)$. For any N let $N_i = \lfloor p_i N \rfloor$ and $M := N - \sum_i N_i \geq 0$. Define the following protocol for ω:

$$\hat{\mathcal{F}}_N = 1^{\otimes M} \otimes (p_1^{-1}\mathcal{E}_1)^{\otimes N_1}\mathcal{F}_{1,N_1} \otimes (p_2^{-1}\mathcal{E}_2)^{\otimes N_2}\mathcal{F}_{2,N_2} \otimes \cdots ,$$

where $1^{\otimes M}$ is the map $M_1(\mathbb{C}) \ni 1 \mapsto 1 \in \mathfrak{A}_A^{\otimes M}$. Because $\omega_N^+ = \omega_M^+ \otimes \omega_{N_1}^+ \otimes \omega_{N_2}^+ \otimes \cdots$ it follows straightforwardly that $\{\hat{\mathcal{F}}_N\}$ is a distillation protocol for ω with rate $\sum_i r_i$. (e4) then follows. $\qquad\square$

Instead of (e5) we have

$$\frac{1}{N}E_D(\omega^{\otimes N}) = E_D(\omega), \tag{3.42}$$

as is obvious from the definition. $E_D(\omega)$ is not in general convex in the sense of property (e3), although it is convex on pure states (this property is not so obvious). Since there are no pure states in the type III case relevant for quantum field theory, this is at any rate not helpful for us, and we have no analogue of (e3).

For us, it is most important that the distillable entropy has the superadditivity property (e6), as remarked without proof e.g. in [34].

Lemma 9 $E_D(\omega)$ *satisfies (e6), i.e. for any state* ω *on* $\mathfrak{A} = (\mathfrak{A}_{A_1} \otimes \mathfrak{A}_{A_2}) \otimes (\mathfrak{A}_{B_1} \otimes \mathfrak{A}_{B_2})$ *the restrictions* ω_i *to* $\mathfrak{A}_{A_i} \otimes \mathfrak{A}_{B_i}$ *satisfy* $E_D(\omega_1) + E_D(\omega_2) \leq E_D(\omega)$.

Proof Let $\{\mathcal{F}_{N,1}\}$ be a distillation protocol for ω_1, i.e. a sequence of cp maps such that $\|\mathcal{F}_{N,1}^*\omega_1^{\otimes N} - \omega_{n_{N,1}}^+\| \to 0$. The rate of this protocol is $\lim(\ln n_{N,1})/N = r_1$, and similarly for ω_2. We claim that $\{\mathcal{F}_{N,1} \otimes \mathcal{F}_{N,2}\}$ is a distillation protocol for ω with rate $r = r_1 + r_2$. Letting $\sigma_N = (\mathcal{F}_{N,1} \otimes \mathcal{F}_{N,2})^*\omega^{\otimes N}$, and $X \in M_{n_{N,1}}(\mathbb{C}) \otimes M_{n_{N,2}}(\mathbb{C})$ and $1_i \equiv 1_{n_{N,i}}$, $P_i^+ \equiv P_{n_{N,i}}^+$, we have

$$|\sigma_N((1_1 - P_1^+) \otimes 1_2 \cdot X)| \leq \sigma_N\left((1_1 - P_1^+) \otimes 1_2 \cdot XX^* \cdot (1_1 - P_1^+) \otimes 1_2\right)^{\frac{1}{2}}$$

$$\leq \|XX^*\|^{\frac{1}{2}}\sigma_N\left((1_1 - P_1^+) \otimes 1_2\right)^{\frac{1}{2}}$$

$$= \|X\| \left|\mathcal{F}_{N,1}^*\omega_1^{\otimes N}(1_1 - P_1^+) - \omega_{n_{N,1}}^+(1_1 - P_1^+)\right|^{\frac{1}{2}}$$

$$\to 0 \quad \text{as } N \to \infty$$

$$\tag{3.43}$$

uniformly for $\|X\| \leq 1$. The same conclusion can be drawn, with similar proof, for $\sigma_N(X \cdot (1_1 - P_1^+) \otimes 1_2)$ and for $(1 \leftrightarrow 2)$. We have

$$\omega^+_{n_{N,1}n_{N,2}}(X) = \omega^+_{n_{N,1}} \otimes \omega^+_{n_{N,2}}(X) = \frac{\sigma_N(P_1^+ \otimes P_2^+ \cdot X \cdot P_1^+ \otimes P_2^+)}{\sigma_N(P_1^+ \otimes P_2^+)}, \qquad (3.44)$$

and therefore in view of (3.43) (and the analogous relations for $(1 \leftrightarrow 2)$),

$$\begin{aligned}
&|(\mathcal{F}_{N,1} \otimes \mathcal{F}_{N,2})^* \omega^{\otimes N}(X) - \omega^+_{n_{N,1}n_{N,2}}(X)| \\
&= |\sigma_N(X) - \omega^+_{n_{N,1}} \otimes \omega^+_{n_{N,2}}(X)| \\
&\leq |\sigma_N(X) - \sigma_N(P_1^+ \otimes P_2^+ \cdot X \cdot P_1^+ \otimes P_2^+)| + |1 - \sigma_N(P_1^+ \otimes P_2^+)^{-1}| \|X\| \to 0
\end{aligned}$$
$$(3.45)$$

uniformly for $\|X\| \leq 1$, which immediately gives the claim. Thus, we see that $\{\mathcal{F}_{N,1} \otimes \mathcal{F}_{N,2}\}$ is a distillation protocol, whose rate is evidently $r_1 + r_2$. Choosing r_i arbitrarily close to $E_D(\omega_i)$, we see that there is a protocol for ω whose rate is at least $E_D(\omega_1) + E_D(\omega_2) - \epsilon$ for any $\epsilon > 0$, which implies superadditivity (e6).   □

It might be guessed from the involved variational characterization of $E_D(\omega)$ that this quantity is difficult to calculate in practice even for the simplest examples, and this expectation turns out to be correct. For us, the usefulness of this quantity lies in the fact it has the very convenient property (e6), and that it is a lower bound for a large class of entanglement measures, in particular the relative entropy of entanglement[9]:

Theorem 5 *For any normal state on $\mathfrak{A} = \mathfrak{A}_A \otimes \mathfrak{A}_B$, and any entanglement measure satisfying (e2), (e4), (e5) and normalization $E(\omega_n^+) = \ln n$, we have $E(\omega) \geq E_D(\omega|\mathfrak{N})$ and in particular $E_R(\omega) \geq E_D(\omega|\mathfrak{N})$, where $\mathfrak{N}$ is any type I subfactor of $\mathfrak{A}$ of the form $\mathfrak{N} = \mathfrak{N}_A \otimes \mathfrak{N}_B$, with $\mathfrak{N}_A \subset \mathfrak{A}_A$, $\mathfrak{N}_B \subset \mathfrak{A}_B$ intermediate type I subfactors.*

Remark 3 The existence of many such intermediate type I subfactors exhausting $\mathfrak{A}$ is guaranteed by the split property.

The proof of this Theorem is given in [25] for the case of finite-dimensional type I algebras, where it is shown more precisely that $\frac{1}{N}E(\omega^{\otimes N}) \geq E_D(\omega) - \epsilon$ for sufficiently large N depending on $\epsilon > 0$. This immediately implies the theorem in view of (e5) for finite-dimensional type I algebras. Inspection of the proof [25] shows that it can be generalized fairly easily to the case of infinite-dimensional type I algebras by a straightforward approximation argument. The general case then follows trivially in view of (e4), $E_R(\omega) \geq E_R(\omega|\mathfrak{N})$.

---

[9]We even get the statement for the "asymptotic" relative entropy of entanglement defined by $E_R^\infty(\omega) = \lim_{n\to\infty} \frac{1}{n}E_R(\omega^{\otimes n})$. Note that the limit exists: Use (e5) and Lemma 12 of [25].

3.8 Summary of Entanglement Measures

We summarize the various entanglement measures and some of their properties and relationships in the following table. Unlisted properties may either be false or unknown to the authors.[10]

Measure	Properties	Relationships	$E(\omega_n^+)$
E_B	(e0), (e2), (e3), (e4)		$\sqrt{2}$
E_D	(e0), (e1), (e2), (e4), (e6)	$E_D \leq E_R, E_N, E_M, E_I$	$\ln n$
E_R	(e0), (e1), (e2), (e3), (e4), (e5)	$E_D \leq E_R \leq E_N, E_M, E_I$	$\ln n$
E_N	(e0), (e1), (e4), (e5)	$E_D, E_R \leq E_N \leq E_M$	$\ln n$
E_M	(e0), $\overline{(e3)}$, (e4), (e5)	$E_D, E_R, E_N \leq E_M$	$\frac{3}{2}\ln n$
E_I	(e0), (e2), (e4), (e5)	$E_D, E_R \leq E_I$	$2\ln n$

References

1. R. Horodecki, P. Horodecki, M. Horodecki, K. Horodecki, Quantum entanglement. Rev. Mod. Phys. **81**(2), 865–942 (2009)
2. G. Vidal, Entanglement monotones. J. Mod. Opt. **47**, 355–376 (2000)
3. V. Vedral, M.B. Plenio, Entanglement measures and purification procedures. Phys. Rev. A. **57**, 3 (1998)
4. V. Vedral, M.B. Plenio, M.A. Rippin, P.L. Knight, Quantifying entanglement. Phys. Rev. Lett. **78**, 2275–2279 (1997)
5. G. Lindblad, Entropy, information, and quantum measurements. Commun. Math. Phys. **33**, 305–322 (1973)
6. W.F. Stinespring, Positive functions on C^*-algebras. Proc. Ame. Math. Soc. **6**, 211–216 (1955)
7. M.A. Nielsen, I.L. Chuang, *Quantum Computation and Quantum Information*, (Cambridge University Press: Cambridge, 2010)
8. G. Giedke, J.I. Cirac, The characterization of Gaussian operations and Distillation of Gaussian States. Phys. Rev. A **66**, 032316 (2002)
9. R. Verch, R.F. Werner, Distillability and positivity of partial transposes in general quantum field systems. Rev. Math. Phys. **17**, 545 (2005)
10. M. Bell, K. Gottfried, M. Veltman, *John S. Bell on the foundations of quantum mechanics* (World Scientific Publishing, Singapore, 2001)
11. J.S. Bell, On the Einstein Podolsky Rosen paradox. Physics **1**, 195 (1964)
12. S.J. Summers, R. Werner, Maximal violation of Bell's inequalities is generic in quantum field theory. Commun. Math. Phys. **110**, 247–259 (1987)
13. S.J. Summers, R. Werner, Maximal violation of Bell's inequalities for algebras of observables in tangent spacetime regions. Ann. Inst. H. Poincaré **49**, 2 (1988)
14. S.J. Summers, R. Werner, Bell's inequalities and quantum field theory. I. General setting. J. Math. Phys. **28**, 2440–2447 (1987)
15. B.S. Tsirelson, Quantum generalizations of Bell's inequality. Lett. Math. Phys. **4**, 93 (1980)
16. J.F. Clauser, M.A. Horne, A. Shimony, R.A. Holt, Phys. Rev. Lett. **49**, 1804 (1969)

---

[10]Property (e4) for E_M has been proven only in a restricted sense, and instead of (e3) we have the opposite: concavity $\overline{(e3)}$.

17. N. Gisin, Hidden quantum nonlocality revealed by local filters. Phys. Lett. A **210**, 151–156 (1996)
18. M.B. Plenio, S. Virmani, An Introduction to entanglement measures. Quant. Inf. Comput. **7**, 1 (2007)
19. J.C. Baez, T. Fritz, A Bayesian characterization of relative entropy. Theor. Appl. Categories **29**, 421–456 (2014)
20. H. Araki, Relative entropy for states of von Neumann algebras. Publ. RIMS Kyoto Univ. **11**, 809–833 (1976)
21. H. Araki, Relative entropy for states of von Neumann algebras II. Publ. RIMS Kyoto Univ. **13**, 173–192 (1977)
22. H. Araki, Relative Hamiltonian for faithful normal states of a von Neumann algebra. Publ. RIMS Kyoto Univ. **9**, 165–209 (1973)
23. M. Ohya, D. Petz, *Quantum Entropy and its Use, Theoretical and Mathematical Physics* (Springer, Berlin, Heidelberg, 1993)
24. A. Uhlmann, Relative entropy and the Wigner-Yanase-Dyson-Lieb concavity in a interpolation theory. Commun. Math. Phys. **54**, 21–32 (1977)
25. M.J. Donald, M. Horodecki, O. Rudolph, Continuity of relative entropy of entanglement. Phys. Lett. A **264**, 257–260 (1999)
26. N. Datta, Min- and max. relative entropies and a new entanglement monotone. IEEE Trans. Inf. Theor. **55**, 2816–2826 (2009)
27. K. Fredenhagen, F.-C. Greipel, *Nuclearity index as a measure for entanglement* Unpublished preprint
28. R.V. Kadison, J.R. Ringrose, *Fundamentals of the Theory of Operator Algebras.* (Academic Press: New York, I (1983), II (1986))
29. K. Fredenhagen, On the modular structure of local algebras of observables. Commun. Math. Phys. **97**, 79–89 (1985)
30. G. Lechner, K. Sanders, Modular nuclearity: a generally covariant perspective. Axioms **5**, 5 (2016)
31. F. Hansen, The fast track to Löwner's theorem. Lin. Alg. Appl. **438**, 4557–4571 (2013)
32. E.A. Carlen, Trace inequalities and entropy: An introductory course. https://www.ueltschi.org/AZschool/notes/EricCarlen.pdf
33. E.M. Rains, Bound on distillable entanglement. Phys. Rev. A **63**, 019902 (2000). Phys. Rev. A 60, 179 (1999); Erratum
34. M.M. Wolf, G. Giedke, J.I. Cirac, Extremality of Gaussian quantum states. Phys. Rev. Lett. **96**, 080502 (2006)

Chapter 4
Upper Bounds for E_R in QFT

Abstract In this chapter we derive some upper bounds on the relative entanglement entropy in quantum field theories, using nuclearity conditions such as the BW-nuclearity or modular nuclearity condition. We consider free fields, 2-dimensional integrable models with factorizing scattering-matrices and CFTs.

In this chapter, we derive some upper bounds on the relative entanglement entropy in quantum field theories. To illustrate the idea, let us first consider the spatial slice $\mathcal{C} = \{t = 0\}$ in $d + 1$-dimensional Minkowski spacetime, and let A and B be two disjoint open regions in $\mathcal{C}$. Let O_A and O_B be the domains of dependence of these regions and $\mathfrak{A}(O_A)$ and $\mathfrak{A}(O_B)$ the associated algebras, see Fig. 1.1. The ground state (vacuum) ω_0 of the QFT gives rise to a GNS-triple $(\pi_0, \mathcal{H}_0, |0\rangle)$, which yields the von Neumann algebras $\mathfrak{A}_A := \pi_0(\mathfrak{A}(O_A))''$ and $\mathfrak{A}_B := \pi_0(\mathfrak{A}(O_B))''$. Note that $\mathfrak{A}_A$ and $\mathfrak{A}_B$ commute, due to causality. We want to investigate when ω_0 defines a normal state on $\mathfrak{A}_A \otimes \mathfrak{A}_B$, i.e. when $\mathfrak{A}_A$ and $\mathfrak{A}_B$ are statistically independent, and what its relative entanglement entropy is.

When the theory satisfies the BW-nuclearity condition (a5), the statistical independence of $\mathfrak{A}_A$ and $\mathfrak{A}_B$ follows from the assumption that the distance $\mathrm{dist}(A, B) > 0$ is positive. Moreover, we will show in Sect. 4.1 that $E_R(\omega_0|_{\mathfrak{A}_A \otimes \mathfrak{A}_B})$ can be estimated in terms of the 1-nuclear norm of the operator Θ which appears in the definition of the BW-nuclearity condition. Using similar methods one can also estimate the relative entanglement entropy of thermal (KMS) states.

The illustrative example above can easily be generalised as follows. We can choose two disjoint regions A and B in a Cauchy surface $\mathcal{C}$ of a globally hyperbolic spacetime $(\mathcal{M}, g)$, and we let O_A and O_B denote their domains of dependence, see Fig. 1.1. We let ω be a state on the QFT on the entire spacetime with GNS-triple $(\pi, \mathcal{H}, |\Omega\rangle)$, and we introduce the von Neumann algebras $\mathfrak{A}_A := \pi(\mathfrak{A}(O_A))''$ and $\mathfrak{A}_B := \pi(\mathfrak{A}(O_B))''$. Once again we can ask whether ω defines a normal state on $\mathfrak{A}_A \otimes \mathfrak{A}_B$ and what its relative entanglement entropy is. In general, however, the BW-nuclearity condition

© The Author(s), under exclusive licence to Springer Nature Switzerland AG 2018   69
S. Hollands and K. Sanders, *Entanglement Measures and Their Properties*
in Quantum Field Theory, SpringerBriefs in Mathematical Physics 34,
https://doi.org/10.1007/978-3-319-94902-4_4

is no longer available, due to the lack of a Hamiltonian operator. For this reason we will consider a modular nuclearity condition instead.

Let us assume for simplicity[1] that $|\Omega\rangle$ is cyclic and separating for $\mathfrak{A}_A$ and for $\mathfrak{A}_B$. (This will be the case e.g. if $\omega = \omega_0$ on Minkowski spacetime, or if ω is any other state with finite energy, by the same argument as in the Reeh-Schlieder theorem. For results in curved spacetimes, see e.g. [2].) In particular, $|\Omega\rangle$ is then cyclic and separating for $\mathfrak{A}_{B'} = \pi(\mathfrak{A}(O_{B'}))''$, where $B' := C \setminus \overline{B}$. Let $\Delta_{B'}$ be the corresponding modular operator, and note that $\mathfrak{A}_{B'} \subset \mathfrak{A}'_B$. Instead of the BW-nuclearity condition (a5), we are going to impose/use the following **modular nuclearity condition**:

(a5') The operator

$$\Phi^A : \mathfrak{A}_A \to \mathcal{H} , \quad \Phi^A(a) = \Delta^{\frac{1}{4}}_{B'} a |\Omega\rangle$$

has $\|\Phi^A\|_1 < \infty$ when dist$(A, B) > 0$ is positive.

As shown in [3], (a5') implies (a5) in Minkowski spacetime without the bounds on the nuclear norms. The nuclearity condition (a5') again suffices to prove the statistical independence of $\mathfrak{A}_A$ and $\mathfrak{A}_B$, and we can estimate the relative entanglement entropy of ω between $\mathfrak{A}_A$ and $\mathfrak{A}_B$ using $\|\Phi^A\|_1$, making use of Theorems 3 and 4. Indeed, noting that $\mathfrak{A}_{B'} \subset \mathfrak{A}'_B$, it follows arguing as in the proof of Proposition 3 that $\|\Psi^A\|_1 \leq \|\Phi^A\|_1$, where Ψ^A is the map (3.26) appearing in the definition of E_M.

In the first section below, we point out some general upper bounds that follow from the BW-nuclearity condition. (The corresponding results for modular operators already follow from the Theorems 3 and 4). In the sections after that we apply the various nuclearity conditions to obtain concrete upper bounds for free fields, 2-dimensional integrable models with factorizing scattering-matrices and CFTs.

4.1   General Upper Bounds From BW-Nuclearity

The first type of general bound is for the ground (vacuum) state, ω_0, and holds for a theory on Minkowski spacetime with mass gap satisfying the BW-nuclearity condition (a5). Our result is:

Theorem 6 *(1) Assume that the Hamiltonian $H = P^0$ in the vacuum representation has a mass gap, spec$(H) \subset \{0\} \cup [m, \infty)$, with $m > 0$. Let A and B be contained in balls of radius r in a $t = 0$ time slice, separated by the distance R. As usual we set $\mathfrak{A}_A = \pi_0(\mathfrak{A}(O_A))''$, where O_A is the causal diamond with base A, and similarly for B. Assume that the BW-nuclearity condition (2.27) holds with constants c, n. Then the relative entanglement entropy in the vacuum state between A and B satisfies, for any $k < 1$, an upper bound of the form*

$$E_R(\omega_0) \lesssim C \exp[-(mR)^k], \tag{4.1}$$

---

[1]A modular operator can still be defined when these assumptions are not met [1], and our estimates, e.g. in the proof of Lemma 3, still hold.

for sufficiently large $R/c \gg 1$, where C is a constant depending on c, k, n.

(2) Under the same assumptions as in (1) but not necessarily $m > 0$, we have for $R/c \ll 1$ an upper bound of the form

$$E_R(\omega_0) \lesssim \sqrt{2} \left(\frac{c \operatorname{ctg} \frac{\pi}{4n}}{R} \right)^n . \tag{4.2}$$

Thus, we see from part (1) that the relative entanglement entropy decays almost exponentially as the separation R between the regions tends to infinity in the presence of a mass gap. Since free bosons [4] and fermions [5] in $d + 1$ dimensions are known to satisfy the BW-nuclearity condition, one immediately gets almost exponential decay in those models. The upper bound expressed by part (2) for short distances shows that the growth of E_R is not faster than an inverse power of R, which depends on the constant n in the nuclearity condition. For free fields, $n = d, c \propto r$, giving thus for $R \ll r$

$$E_R(\omega_0) \lesssim c_d \left(\frac{r}{R} \right)^d , \tag{4.3}$$

so this upper bound falls short of the expected "area law". For a better bound qualitatively consistent with the area law in the case of Dirac fields, see Sect. 4.2.2.

The second theorem is concerned specifically with thermal states (KMS states) ω_β. Again, we take A and B as balls of radius r separated by the distance R in some slice $\mathbb{R}^d$ of Minkowski spacetime for simplicity. We go to the GNS-representation $(\mathcal{H}_\beta, \pi_\beta, |\Omega_\beta\rangle)$ of ω_β, in which this state is represented by the vector $|\Omega_\beta\rangle$. The state ω_β is not pure and the representation is highly reducible. The time-translation subgroup is implemented by the unitary $U_\beta(t) = e^{itH_\beta}$. Unlike in the vacuum representation, the generator H_β always has as its spectrum the entire real line $\operatorname{spec}(H_\beta) = \mathbb{R}$, even if the theory has a mass gap in the vacuum sector. As a replacement of the BW-nuclearity condition in this case, we assume that (1) the point $\{0\}$ is a non-degenerate eigenvalue of H_β associated with the vector $|\Omega_\beta\rangle$, (2) Letting P_+ be the spectral projection of H_β corresponding to the set $(0, \infty)$ and P_- that corresponding to the set $(-\infty, 0)$, the maps $\Theta_z^\pm : \mathfrak{A}_B \to \mathcal{H}_\beta$

$$\Theta_{r,z}^\pm(b) := P_\pm e^{\pm izH_\beta} b|\Omega_\beta\rangle , \qquad \Im(z) > 0, \tag{4.4}$$

are assumed to be nuclear. By invariance under spatial translations, the same then obviously holds for the region A.

Theorem 7 *Assume that A and B are as in Theorem 6 and assume*

$$\|\Theta_{r,z}^\pm\|_1 \leq |\Im(z)|^{-\alpha} \exp(c/|\Im(z)|)^n \tag{4.5}$$

for a constant c depending on r, β and $n > 0, \alpha > 1$. Then

$$E_R(\omega_\beta) \leq C R^{-\alpha+1} \tag{4.6}$$

for sufficiently large $R = dist(A, B)$ and a constant C depending on r, β.

The theorem leaves the possibility that the relative entanglement entropy of two regions in a thermal state may decay more slowly than that in a vacuum state. The precise rate of the upper bound is related to spectral information, which in our case is encoded in the assumption about the nuclear norms. The last theorem concerns massless theories.

Theorem 8 *In a massless theory on Minkowski spacetime with vacuum ω_0, assume that the 1-norm of the map $\Theta_{r,z} : a \mapsto e^{izH} a |0\rangle$ fulfills (4.5) for $\Im z > 0$. Then*

$$E_R(\omega_0) \leq C R^{-\alpha} \tag{4.7}$$

for sufficiently large $R = dist(A, B)$ and a constant C depending on r.

The strategy to prove these theorems is as follows: we show that we can write the relevant state ω on $\mathfrak{A}_A \otimes \mathfrak{A}_B$ in the form $\omega = \sum_j \phi_j \otimes \psi_j$, with normal linear functionals ϕ_j and ψ_j satisfying a bound on $\nu := \sum_j \|\phi_j\| \cdot \|\psi_j\|$, which is controlled by the respective nuclearity condition assumed in each theorem. The desired bound on $E_R(\omega)$ is then obtained in conjunction with Lemma 3 and Theorem 3, which imply that

$$E_R(\omega) \leq \ln \nu. \tag{4.8}$$

We now turn to the detailed estimation in each case.

4.1.1 Proof of Theorem 6

The basic idea of all three proofs is to consider suitable correlation functions $f_{a,b}(z)$ whose analyticity properties in z encode the commuting nature of $\mathfrak{A}_A$ and $\mathfrak{A}_B$. Part (1) of the proof of the present theorem relies on an argument due to [6, 7] (up to a slightly better control on the bounds), which we repeat here merely for the convenience of the reader and to set the stage for part (2).

Proof of Theorem 6, Part (1): Following [6, 7], we consider the function

$$f_{a,b}(z) = \begin{cases} \langle 0|a P e^{+izH} b|0\rangle & \text{for } 0 < \Im(z), \\ \langle 0|b P e^{-izH} a|0\rangle & \text{for } 0 > \Im(z), \end{cases} \tag{4.9}$$

where $a \in \mathfrak{A}_A, b \in \mathfrak{A}_B$, and where $P = 1 - |0\rangle\langle 0|$. $z \mapsto f_{a,b}(z)$ is by construction analytic for $\Im(z) \neq 0$, i.e. away from the real axis. For the jump across the real axis, one finds

$$f_{a,b}(t + i0) - f_{a,b}(t - i0) = \langle 0|[a, e^{itH} b e^{-itH}]|0\rangle = 0 \tag{4.10}$$

as long as $|t| < R$, because then the time-translated region O_B (by t) remains spacelike to O_A. By the edge-of-the-wedge theorem (see Appendix A.1), the function

Fig. 4.1 The image of the
contour C_ρ for two different
values of ρ

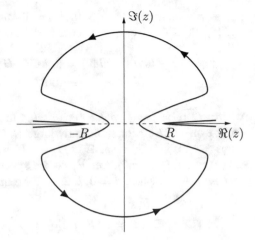

$f_{a,b}(z)$ may thus be extended to an analytic function in the doubly cut plane $\mathbb{C} \setminus \{z \in \mathbb{C} \mid \Im(z) = 0, |\Re(z)| \geq R\}$. We consider next for $|t| < R$ the mapping $w \mapsto z = 2t/(w + w^{-1})$ which maps the open disk $\{|w| < 1\}$ into the doubly cut plane. The image of the contour $C_\rho : \varphi \mapsto \rho e^{i\varphi}, 0 < \rho < 1$ in the doubly cut z-plane under this mapping is illustrated in the Fig. 4.1. Applying Cauchy's formula to this contour gives

$$\langle 0|ab|0\rangle - \langle 0|a|0\rangle \langle 0|b|0\rangle = \int_{C_\rho} \frac{dw}{2\pi i\; w}\, f_{a,b}\left(\frac{2t}{w + w^{-1}}\right)$$
$$\rightarrow \frac{1}{2\pi} \int_0^{2\pi} d\varphi\, f_{a,b}\left(\frac{t}{\cos\varphi}\right), \tag{4.11}$$

sending $\rho \rightarrow 1$ in the second line.

For the part above resp. below the real axis we can use the respective representations of $f_{a,b}$, and this gives

$$\langle 0|ab|0\rangle - \langle 0|a|0\rangle \langle 0|b|0\rangle = \frac{1}{2\pi} \int_0^\pi d\varphi\, \langle 0|a\, P \exp(+it H/\cos\varphi) b|0\rangle$$
$$+ \frac{1}{2\pi} \int_\pi^{2\pi} d\varphi\, \langle 0|b\, P \exp(-it H/\cos\varphi) a|0\rangle \tag{4.12}$$

for all $|t| < R$. Following [7], we next proceed as follows. We multiply this identity with a smooth test function of the form $g_k(t/R)/R$ where the support of $g_k(t)$ is contained in $(-1, 1)$. For convenience, we normalize $g_k(t)$ so that $\int_{\mathbb{R}} g_k(t)dt = 1$ and furthermore make a choice such that $|\tilde{g}_k(E)| \leq C_k \exp(-|E|^k)$, for $k < 1$ for

$|E| \to \infty$, and for some $C_k > 0$. It is well-known that such a choice is possible, see e.g. [8, 9].[2] This gives a representation of the form

$$\langle 0|ab|0\rangle - \langle 0|a|0\rangle \langle 0|b|0\rangle = \langle 0|aG(-H)Pb|0\rangle + \langle 0|bG(H)Pa|0\rangle$$
$$= \langle 0|b\Xi^+(a)|0\rangle + \langle 0|a\Xi^-(b)|0\rangle, \tag{4.14}$$

where $\sqrt{2\pi}G(E) = \int_0^\pi \tilde{g}(RE/\cos\varphi)\,d\varphi$. In particular, it follows that $|G(E)|$ is of order $\exp(-|RE|^k)$ for sufficiently large $|RE|$. The terms on the right side define maps $\Xi^+ : \mathfrak{A}_A \to \mathcal{H}$ resp. $\Xi^- : \mathfrak{A}_B \to \mathcal{H}$ by $\Xi^+(a) = PG(H)a|0\rangle$ respectively $\Xi^-(b) = PG(-H)b|0\rangle$. Let E_j be the spectral projector for H onto the interval $[mj, m(j+1))$, where $j = 0, 1, 2, \ldots$. We may then write

$$\Xi^+(a) = G(H)Pa|0\rangle = \sum_{j=1}^\infty e^{\beta_j H} G(H) E_j \Theta_{\beta_j,r}(a), \tag{4.15}$$

where the $\beta_j > 0$ are to be chosen. (There is no $j = 0$ term in the sum because P projects out the vacuum state and H has a mass gap.) Using this formula, we estimate:

$$\|\Xi^+\|_1 \le \sum_{j=1}^\infty \|G(H)E_j\| e^{\beta_j(j+1)m} \|\Theta_{\beta_j,r}\|_1$$
$$\le C_k \sum_{j=1}^\infty e^{-(Rmj)^k} e^{\beta_j m(j+1)} e^{(c/\beta_j)^n} \tag{4.16}$$
$$\le C_k \sum_{j=1}^\infty e^{-(Rmj)^k} e^{3(cmj)^{n/(n+1)}}$$

using the properties of G and the nuclearity assumption to go to line two, and making the choice $\beta_j = c(cmj)^{-1/(n+1)}$ to go to line three. Choosing any $k > n/(n+1)$, there follows the bound $\|\Xi^+\|_1 \lesssim Ce^{-(mR)^k}$ (for a new constant $C = C(k,n,c)$), and we can get the same type of estimate for Ξ^-. Combining these, we find that there exist functionals ψ_j, φ_j on $\mathfrak{A}(O_B), \mathfrak{A}(O_A)$ respectively such that

$$\omega_0(ab) = \omega_0(a)\omega_0(b) + \sum_{j=1}^\infty \varphi_j(a)\psi_j(b) \tag{4.17}$$

---

[2]A permissible non-normalized choice of $g_k(t)$ is for instance $g_k(t) = h_k(1+t)h_k(1-t)$, where $h_k(t)$ has the Fourier Laplace transform $\int_0^\infty h_k(t)e^{-Et}\,dt = e^{-E^k}$. For further discussion and explicit formulas, see e.g. [10], where one finds

$$h_k(t) = -\frac{1}{\pi} \sum_{j\ge 1} \frac{(-1)^j \Gamma(jk+1)\sin(\pi jk)}{j!t^{jk+1}}. \tag{4.13}$$

Fig. 4.2 The contour C_R

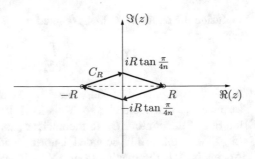

and such that $\sum_j \|\varphi_j\| \, \|\psi_j\| \leq Ce^{-(mR)^k}$ for large R. Thus, we are in the situation of Lemma 3 with $\nu = 1 + Ce^{-(mR)^k}$. Lemma 3 and Theorem 3 together with the elementary bound $\ln(1 + x) \leq x$ now implies the statement.

Proof of Theorem 6, Part (2): We view $f_{a,b}(z)$ as defining a function Ξ_z from the doubly cut plane $\mathbb{C} \setminus \{z \in \mathbb{C} \mid \Im(z) = 0, |\Re(z)| \geq R\}$ into the linear maps $\mathfrak{B} = \mathfrak{B}(\mathfrak{A}_B, (\mathfrak{A}_A)_*)$ by the formula $[\Xi_z(b)](a) = f_{a,b}(z)$. The BW-nuclearity assumption implies the nuclearity of this map for $|\Im z| > 0$, with 1-norm bounded from above by $\|\Xi_z\|_1 \leq e^{(c/|\Im z|)^n}$. Without loss of generality, we may assume that $n > 1/2$, since increasing n makes the bound less tight (in practice, one expects $n \sim d$ anyhow).

Let C_R be the contour shown in Fig. 4.2. Consider the map $z \mapsto \Upsilon_z$ from the interior of C_R to $\mathfrak{B}$ defined by

$$\Upsilon_z = \varphi(z) \cdot \Xi_z \tag{4.18}$$

where

$$\varphi(z) = C(n, R, c) \exp\left[-\sqrt{2}\left(\frac{c}{(R + z)\sin\frac{\pi}{4n}}\right)^n\right] \exp\left[-\sqrt{2}\left(\frac{c}{(R - z)\sin\frac{\pi}{4n}}\right)^n\right], \tag{4.19}$$

and where

$$C(R, n, c) = \exp\left[2\sqrt{2}\left(\frac{c}{R\sin\frac{\pi}{4n}}\right)^n\right].$$

Since C_R is contained in the doubly cut plane $\mathbb{C} \setminus \{\Im(z) = 0, |\Re(z)| \geq R\}$ where Ξ_z is analytic, and since φ is analytic in the interior of C_R it follows that $z \mapsto \Upsilon_z$ is holomorphic in the interior of C_R. Using the explicit form of the function $\varphi(z)$ and the fact that $\|\Xi_z\|_1 \leq e^{(c/|\Im z|)^n}$ in particular along the contour C_R, we see that

$$\|\Upsilon_z\|_1 \leq |\varphi(z)| e^{(c/|\Im z|)^n} \leq \exp\left[\sqrt{2}\left(\frac{c\,\mathrm{ctg}\frac{\pi}{4n}}{R}\right)^n\right] \quad \text{for all } z \in C_R. \tag{4.20}$$

Also, by definition, $\Xi_0 = \Upsilon_0$.

Inside the contour, Υ_z has, for fixed $a \in \mathfrak{A}_A, b \in \mathfrak{A}_B$, the contour integral representation

$$[\Upsilon_z(b)](a) = \int_{C_R} \frac{dw}{2\pi i \, (w-z)}[\Upsilon_w(b)](a). \tag{4.21}$$

Making an argument of the same kind as in [11] based on this contour integral formula, we can see that $z \mapsto \Upsilon_z$ is in fact a strongly holomorphic map from the interior of the contour C_R to the nuclear (not just bounded) operators from $\mathfrak{A}_B \to (\mathfrak{A}_A)_*$ with uniformly bounded 1-norm inside the contour C_R. By Lemma 14 in Appendix A.1, the maximum of $\|\Upsilon_z\|_1$ is achieved on the boundary C_R, and thus $\|\Upsilon_z\|_1 \leq \exp[\sqrt{2}(\frac{c \, \mathrm{ctg} \frac{\pi}{4n}}{R})^n]$ also for z inside—and not just on—the contour C_R. Thus, from $\Xi_0 = \Upsilon_0$, we conclude in particular that $\|\Xi_0\|_1 \leq \exp[\sqrt{2}(\frac{c \, \mathrm{ctg} \frac{\pi}{4n}}{R})^n]$. Arguing now as in Part (1), we get (4.17) with $\sum_j \|\varphi_j\| \, \|\psi_j\| \leq \exp[\sqrt{2}(\frac{c \, \mathrm{ctg} \frac{\pi}{4n}}{R})^n]$ which holds for all R, in particular for $R/c \ll 1$. Thus, we are in the situation of Lemma 3 with $\nu = 1 + \exp[\sqrt{2}(\frac{c \, \mathrm{ctg} \frac{\pi}{4n}}{R})^n]$.

Lemma 3 and Theorem 3 imply that $E_R(\omega) \leq \ln \nu$, which gives the statement. $\qquad \square$

4.1.2   Proof of Theorems 7 and 8

Proof of Theorem 7: As in the previous proof, we omit the reference to the representation π_β and simply write a for $\pi_\beta(a)$. We also write J for the modular conjugation associated with[3] $\mathfrak{A}$ (the v. Neumann closure of $\cup_O \mathfrak{A}(O)$)—*not* with $\mathfrak{A}_A$—and the state ω_β, acting on the GNS-Hilbert space with implementing vector $|\Omega_\beta\rangle$. Fix $a \in \mathfrak{A}_A \vee J\mathfrak{A}_A J, b \in \mathfrak{A}_B$, and define the function $F_{a,b}(z)$ as

$$F_{a,b}(z) = \begin{cases} \langle \Omega_\beta | a P_+ e^{izH_\beta} b \Omega_\beta \rangle - \langle \Omega_\beta | b P_- e^{-izH_\beta} a \Omega_\beta \rangle & \text{for } 0 < \Im(z) \\ -\langle \Omega_\beta | a P_- e^{izH_\beta} b \Omega_\beta \rangle + \langle \Omega_\beta | b P_+ e^{-izH_\beta} a \Omega_\beta \rangle & \text{for } 0 > \Im(z), \end{cases} \tag{4.22}$$

which coincides with a function considered also by [12]. Here H_β is the generator of time translations in the representation π_β ("Liouvillean"). For $-R < t < R$ one finds for the jump across the real axis:

$$F_{a,b}(t+i0) - F_{a,b}(t-i0) = \langle \Omega_\beta | [a, e^{itH_\beta} b e^{-itH_\beta}] \Omega_\beta \rangle = 0 \tag{4.23}$$

using space-like commutativity and the fact that the time-translated region O_B remains space-like to O_A as long as the time translation parameter t stays in the range $t \in (-R, R)$. By the edge-of-the-wedge theorem (see Appendix A.1), $F_{a,b}(z)$ therefore defines an analytic function in the doubly cut plane $\mathbb{C} \setminus \{\Im(z) = 0, |\Re(z)| \geq R\}$. Using N times the KMS-condition (applying the last item in Proposition 2 to the pair $\mathfrak{A}, \omega_\beta$) for ω_β, one can derive the "image sum" formula

---

[3]Note that $\pi_\beta(\mathfrak{A})$ has a non-trivial commutant since the the representation π_β is not irreducible.

$$\langle \Omega_\beta | ab\Omega_\beta \rangle - \langle \Omega_\beta | a\Omega_\beta \rangle \langle \Omega_\beta | b\Omega_\beta \rangle = \sum_{k=-N}^{N} F_{a,b}(i\beta k) + \sum_{k=-(N-\frac{1}{2})}^{N-\frac{1}{2}} F_{Ja^*J,b}(i\beta k)$$
$$+ \langle \Omega_\beta | b P_- e^{\beta N H_\beta} a\Omega_\beta \rangle + \langle \Omega_\beta | a P_- e^{\beta N H_\beta} b\Omega_\beta \rangle. \tag{4.24}$$

Next, we view $F_{a,b}(z)$ as defining a function Ξ_z from the doubly cut plane into the linear maps $\mathfrak{B} = \mathfrak{B}(\mathfrak{A}_B, (\mathfrak{A}_A \vee J\mathfrak{A}_A J)_*)$ by the formula $[\Xi_z(b)](a) = F_{a,b}(z)$. The definition of $F_{a,b}(z)$, implies that for $|\Im(z)| > 0$, the nuclear 1-norm of Ξ_z satisfies the upper bound

$$\|\Xi_z\|_1 \leq \|\Theta_{r,z}^+\|_1 + \|\Theta_{r,z}^-\|_1 \leq 2|\Im(z)|^{-\alpha} \exp(c/|\Im(z)|)^n \tag{4.25}$$

using in the second step the assumption of the theorem.

We also consider the map $z \mapsto \Upsilon_z$ from the doubly cut plane $\mathbb{C} \setminus \{\Im(z) = 0, |\Re(z)| \geq R\}$ to $\mathfrak{B}$ defined by

$$\Upsilon_z = R^{-\alpha}(R - z)^\alpha (R + z)^\alpha \cdot \varphi(z) \cdot \Xi_z \tag{4.26}$$

where $\alpha > 0$ is as in the assumptions of the theorem, and where $\varphi(z)$ is the function defined previously in Eq. (4.19). It follows that also $z \mapsto \Upsilon_z$ is strongly analytic as a map from the doubly cut plane to $\mathfrak{B}$.

Now let C_R be the contour of Fig. 4.2, which is entirely within the doubly cut plane. Inside this contour, Υ_z has, for fixed $a \in \mathfrak{A}_A \vee J\mathfrak{A}_A J$, $b \in \mathfrak{A}_B$, a contour integral representation as in (4.21). Using the definition of $\varphi(z)$ and our previous bound on $\|\Xi_z\|_1$, we find that the nuclear 1-norm of Υ_z is bounded by

$$\|\Upsilon_z\|_1 \leq R^{-\alpha} \sup_{w \in C_R} \{|w - R|^\alpha |w + R|^\alpha |\varphi(w)| \cdot \|\Xi_w\|_1\}$$
$$\leq 2 \exp[\sqrt{2}(\tfrac{c \operatorname{ctg} \frac{\pi}{4n}}{R})^n] (\tfrac{1}{2} \sin \tfrac{\pi}{4n})^{-\alpha} \tag{4.27}$$

for all z on the contour C_R. Making again an argument of the same kind as in [11] based on the contour integral formula, we can see that $\|\Upsilon_z\|_1$ must remain bounded also *inside* the contour C_R, and that $z \mapsto \Upsilon_z$ is in fact a holomorphic map from the interior of the contour C_R to the nuclear (not just bounded) operators from $\mathfrak{A}_B \to (\mathfrak{A}_A \vee J\mathfrak{A}_A J)_*$ with uniformly bounded 1-norm inside the contour C_R.

By Lemma 14 in Appendix A.1 the map $z \mapsto \|\Upsilon_z\|_1$ from the interior of C_R in fact assumes its maximum on the boundary, C_R, so (4.27) also holds for z inside C_R. For $k \in \tfrac{1}{2}\mathbb{Z}$ and $|k| \leq N := \lfloor \beta^{-1} R \sin \tfrac{\pi}{4n} \rfloor$, the points $i\beta k$ are inside the contour C_R. Going back from Υ_z to Ξ_z, it follows that, for such k, and for $R/c \gg 1$,

$$\|\Xi_{i\beta k}\|_1 \leq C_1 R^\alpha (R^2 + (\beta k)^2)^{-\alpha} \tag{4.28}$$

for some constant C_1. Therefore

$$\sum_{k\in\frac{1}{2}\mathbb{Z},|k|\leq N} \|\Xi_{i\beta k}\|_1 \leq C_2 R^{-\alpha+1} \tag{4.29}$$

using our assumption that $\alpha > 1$. On the other hand, using (4.24) and our definitions of Ξ_z, $\Theta_z^{\pm}$, we may write

$$\langle\Omega_\beta|ab\Omega_\beta\rangle - \langle\Omega_\beta|a\Omega_\beta\rangle\langle\Omega_\beta|b\Omega_\beta\rangle$$

$$= \sum_{k\in\mathbb{Z},|k|\leq N} [\Xi_{i\beta k}(b)](a) + \sum_{k\in\mathbb{Z}+\frac{1}{2},|k|\leq N} [\Xi_{i\beta k}(b)](Ja^*J) \tag{4.30}$$

$$+ \langle\Omega_\beta|b\Theta_{\beta N,r}^-(a)\Omega_\beta\rangle + \langle\Omega_\beta|a\Theta_{\beta N,r}^-(b)\Omega_\beta\rangle.$$

As a consequence of the nuclearity bound (4.29), the bound $\|\Theta_{\beta N,r}^{\pm}\|_1 \leq (N\beta)^{-\alpha}\exp[(c/N\beta)^n]$ assumed in the theorem, and since $a \mapsto Ja^*J$ is bounded, we can write

$$\langle\Omega_\beta|ab\Omega_\beta\rangle - \langle\Omega_\beta|a\Omega_\beta\rangle\langle\Omega_\beta|b\Omega_\beta\rangle = \sum_j \varphi_j(a)\psi_j(b), \tag{4.31}$$

where the linear functionals φ_j on $\mathfrak{A}_A$ and ψ_j on $\mathfrak{A}_B$ satisfy $\sum_j \|\varphi_j\| \|\psi_j\| \leq C_3 R^{-\alpha+1}$, for a new constant C_3. Thus, we can apply the Lemma 4 and Theorem 3 with $\nu := 1 + C_3 R^{-\alpha+1}$. The bound (4.8) together with $\ln(1+x) \leq x$ then gives the statement. $\square$

Proof of Theorem 8: The proof is very similar to that of the previous theorem and is now based on the correlation function (4.9). We omit the details. $\square$

4.2   Upper Bounds for Free Quantum Field Theories in $d+1$ Dimensions

4.2.1   Free Scalar Fields

The BW-nuclearity condition is well-established for free scalar fields in $d+1$ dimensional Minkowski spacetime both in the massive ($m > 0$) and the massless ($m = 0$) case. Thus, Theorem 6 respectively Theorem 8 apply and provide upper bounds on the entanglement entropy $E_R(\omega_0)$ of two diamonds of size r separated by a distance R.

In fact, in the massive case, an explicit nuclearity bound is given in [4].[4] Therefore Theorem 6 can be applied to infer sub-exponential decay of the entanglement entropy for large separation R. In the massless case, it can be extracted from the work of [13]

---

[4]To be specific, [4] show that in the massive case, $\|\Theta_{\beta,r}\|_1 \leq \exp[(c/\beta)^d|\ln(1-e^{-m\beta/2})|]$ for $r > m^{-1}$ and $0 < \beta \leq r$, where $c = c_0 r$.

that $\|\Theta_{\beta,r}\|_1 \leq \beta^{-(d-2)} \exp(c/\beta)^d$ for $\beta > 0$, $r > 0$ and $d > 2$. Applying Theorem 8 then immediately gives

Proposition 6 *Let ω_0 be the vacuum state of a free $m = 0$ Klein-Gordon field in Minkowski spacetime $\mathbb{R}^{d,1}$. Let $A, B \subset \mathbb{R}^d$ be two balls of radius r separated by a distance R. Then*

$$E_R(\omega_0) \lesssim C R^{-(d-2)}$$

when $R/r \gg 1$, where C is a constant depending on r and $d > 2$.

This upper bound is consistent with a general bound for conformal quantum field theories given below in the remark after Theorem 15. More general, and somewhat better bounds, can be obtained from the modular nuclearity bound $E_R(\omega) \leq E_M(\omega)$ in Theorems 3, and 4 using results of [1] in the case $m > 0$. We formulate them first in the general setting of open regions A and B in a Cauchy surface $\mathcal{C}$ of a globally hyperbolic spacetime $(\mathcal{M}, g)$. Furthermore, we let ω be any quasi-free state on the Weyl algebra of a free scalar field on $(\mathcal{M}, g)$.

Recall from Sect. 2.4.1 that the space of real initial data $K_{\mathbb{R}} := C_0^\infty(\mathcal{C}, \mathbb{R}) \oplus C_0^\infty(\mathcal{C}, \mathbb{R})$ carries a symplectic form

$$\sigma((f_0, f_1), (h_0, h_1)) = \int_{\mathcal{C}} (f_0 h_1 - f_1 h_0) \mathrm{d}V.$$

Following the notations of Sect. 2.2.1, the quasi-free state ω corresponds to an inner product μ on $K_{\mathbb{R}}$, and we let $\mathrm{clo}_\mu K$ denote the complex Hilbert space completion of the complexification K of $K_{\mathbb{R}}$. We let Γ denote the complex conjugation on $\mathrm{clo}_\mu K$ and Σ the operator which implements the symplectic form. The one-particle Hilbert space can then be identified with $\mathcal{H}_1 := \overline{\mathrm{ran}(1 + \Sigma)}$, whereas the full GNS-representation space is the bosonic Fock space $\mathcal{F}_+(\mathcal{H}_1)$ with GNS vector $|\Omega\rangle$.

If $V \subset \mathcal{C}$ is any open region, the real initial data in V generate a symplectic subspace $K_{\mathbb{R}}(V)$ of $K_{\mathbb{R}}$, and the complex initial data in V generate a closed subspace $\mathcal{K}(V) \subset \mathrm{clo}_\mu K$. Let P_V be the orthogonal projection onto $\mathcal{K}(V)$ and note that $\Sigma_V := P_V \Sigma P_V$ implements the symplectic form on $\mathcal{K}(V)$. The local Weyl algebra $\mathfrak{W}(K_{\mathbb{R}}(V), \sigma)$ gives rise to a von Neumann algebra $\mathfrak{A}_V := \pi_\mu(\mathfrak{W}(K_{\mathbb{R}}, \sigma))''$, and we may associate to $(\mathfrak{A}_V, \Omega)$ a modular operator Δ_V and a modular conjugation J_V. These operators are the second quantizations of operators δ_V and j_V on $\mathcal{H}_1$, which we now describe.

First we let $\mathcal{H}_1(V) := \overline{\mathrm{ran}(P_V + \Sigma_V)}$ denote the one-particle space corresponding to the restriction of ω to $\mathfrak{A}_V$, and we note that $\mathcal{H}_1(V) \subset \mathcal{K}(V)$. Then we let $\tilde{R}_V$ be the projection onto $\overline{\mathrm{ran}(P_V - |\Sigma_V|)}$, which is a subspace of $\mathcal{H}_1(V)$. Recall from Sect. 2.2.1 that Ω is separating for $\mathfrak{A}_V$ if and only if $\tilde{R}_V = P_V$, which also entails that $\mathcal{H}_1(V) = \mathcal{K}(V)$. Using these projections we may write the one-particle modular operator $\tilde{\delta}_V$ and modular conjugation $\tilde{j}_V$ on $\mathcal{H}_1(V)$ as

$$\tilde{\delta}_V = \frac{P_V - \Sigma_V}{P_V + \Sigma_V}\,\tilde{R}_V$$

$$\tilde{j}_V = \Gamma\tilde{R}_V.$$

One easily verifies that $\tilde{j}_V\tilde{\delta}_V^{\frac{1}{2}} = s_V$, where $s_V\,\tilde{R}_V\sqrt{P_V + \Sigma_V}\,F = \tilde{R}_V\sqrt{P_V + \Sigma_V}\,\Gamma F$ for complex initial data F in V.

Note that $\mathcal{H}_1(V)$ is isometric to a subspace of $\mathcal{H}_1$. Restricting the domain of this isometry to $\tilde{R}_V\mathcal{H}_1(V)$, it takes the form

$$U_V = \sqrt{1 + \Sigma}\,(P_V + \Sigma_V)^{-\frac{1}{2}}\tilde{R}_V. \tag{4.32}$$

This satisfies $U_V^* U_V = \tilde{R}_V$ and we denote the range projection of U_V by $R_V :=$ $U_V U_V^* = U_V \tilde{R}_V U_V^*$. We may then introduce the modular operator δ_V and the modular conjugation j_V on $\mathcal{H}_1(V)$, defined by

$$\delta_V := U_V\tilde{\delta}_V U_V^* = \sqrt{1+\Sigma}\,\frac{P_V - \Sigma_V}{(P_V + \Sigma_V)^2}\,\tilde{R}_V\sqrt{1+\Sigma} \tag{4.33}$$

$$j_V := U_V\tilde{j}_V U_V^* = \Gamma\sqrt{1-\Sigma}\,(P_V - \Sigma_V^2)^{-\frac{1}{2}}\tilde{R}_V\sqrt{1+\Sigma}.$$

Note that $\delta_V R_V = \delta_V$ and that

$$\delta_V^{\frac{1}{2}} := U\tilde{\delta}_V^{\frac{1}{2}}U^* = \sqrt{1+\Sigma}\,\frac{(P_V - \Sigma_V)^{\frac{1}{2}}}{(P_V + \Sigma_V)^{\frac{3}{2}}}\,\tilde{R}_V\sqrt{1+\Sigma}.$$

We now apply these notations to the case $V = B' = \mathcal{C}\setminus\overline{B}$, i.e. we consider the modular operator $\Delta_{B'}$ associated to the algebra $\mathfrak{A}_{B'}$. The 1-nuclear norm of the operator $\Phi^A : \mathfrak{A}_A \to \mathcal{H}$, $\Phi^A(a) = \Delta_{B'}^{\frac{1}{4}}a|\Omega\rangle$ as in (a5$'$) can then be estimated in terms of one-particle operators as follows:

Theorem 9 *For open regions $A \subset B'$ in $\mathcal{C}$ we define*

$$\alpha := (1 + \delta_A)^{-\frac{1}{2}}R_A\delta_{B'}^{\frac{1}{2}}R_A(1 + \delta_A)^{-\frac{1}{2}}$$

$$= \frac{1}{2}U_A(1 + \Sigma)\frac{(P_{B'} - \Sigma_{B'})^{\frac{1}{2}}}{(P_{B'} + \Sigma_{B'})^{\frac{3}{2}}}\,\tilde{R}_{B'}(1 + \Sigma)U_A^*.$$

Then $\|\Phi^A\|_1 \leq \det(1 - \sqrt{\alpha + j_A\alpha j_A})^{-4}$ *and*

$$E_R(\omega) \leq \ln(\|\Phi^A\|_1) \leq -4\,\mathrm{Tr}\,\ln(1 - \sqrt{\alpha + j_A\alpha j_A}).$$

Proof The equality of both formulae for α can be verified using Eqs. ((4.32) and 4.33) the fact that U_A intertwines the functional calculus of δ_A and $\tilde{\delta}_A$, which yields in particular that

$$(1 + \delta_A)^{-\frac{1}{2}} R_A = \frac{1}{\sqrt{2}} U_A \sqrt{1 + \Sigma}.$$

The main part of the proof is then a particular case of Theorem 3.3 and of Eq. (3.15) of [1] (the latter is originally due to [13]). Instead of reproducing the entire argument, we only give the following remarks. If ω is not cyclic and separating for $\mathfrak{A}_{B'}$, the modular operator $\delta_{B'}$ and conjugation $j_{B'}$ vanish on a certain subspace. However, we may extend $j_{B'}$ to a complex conjugation $\Gamma_{B'}$ on the entire Hilbert space $\mathcal{H}_1$, and we use this conjugation in the doubling procedure in the proof of Theorem 3.3 in [1]. On $\mathcal{H}_1 \oplus \mathcal{H}_1$ we then use the one-particle operator $X_1 := \delta_{B'}^{\frac{1}{4}} \oplus \delta_{B'}^{-\frac{1}{4}}$.

Let H_A be the closed real-linear subspace in $\mathcal{H}_1$ generated by $\sqrt{1 + \Sigma} F$ with real initial data F supported in A, and let H_A° be its symplectic complement. The complex subspaces $\underline{\mathcal{K}}_{\pm}$ constructed in [1] are then the sums $\underline{\mathcal{K}}_{\pm} = \underline{K}_0 \oplus \underline{\mathcal{K}}'_{\pm}$ of the spaces

$$\underline{K}_0 = (H_A^{\circ})^{\perp} \oplus \Gamma_{B'} (H_A^{\circ})^{\perp}$$

$$\underline{K}'_{\pm} = \left\{ \begin{pmatrix} h_1 + ih_2 \\ \pm \Gamma_{B'}(h_1 - ih_2) \end{pmatrix} \mid h_1, h_2 \in R_A H_A \right\}.$$

Because these summands are orthogonal, the orthogonal projections $E_{\pm}$ onto $\underline{\mathcal{K}}_{\pm}$ can be decomposed into sums of orthogonal projections, $E_0 + E'_{\pm}$, where E_0 projects onto $\underline{K}_0$ and $E'_{\pm}$ onto $\underline{K}'_{\pm}$. We then have $X_1 E_{\pm} = X_1 E'_{\pm}$, and one may compute that

$$E'_{\pm} = \begin{pmatrix} (1 + \delta_A)^{-1} R_A & \pm (1 + \delta_A)^{-1} \delta_A^{\frac{1}{2}} j_A \Gamma_{B'} \\ \pm \Gamma_{B'} j_A (1 + \delta_A)^{-1} \delta_A^{\frac{1}{2}} & \Gamma_{B'} (1 + \delta_A)^{-1} \Gamma_{B'} \end{pmatrix}.$$

It turns out that $|X_1 E_{\pm}| = |X_1 E'_{\pm}|$ is unitarily equivalent to the matrix

$$\begin{pmatrix} \alpha + j_A \alpha j_A & 0 \\ 0 & 0 \end{pmatrix}$$

for both signs. The estimate (2.24) of [13] then yields $\| \Phi^A \|_1 \leq \det(1 - \sqrt{\alpha + j_A \alpha j_A})^{-4}$ (cf. (3.15) in [1]), and the estimate on $E_R(\omega)$ follows from our Theorems 3 and 4, together with the well-known formula $\ln \det(X) = \mathrm{Tr} \ln(X)$ for positive operators X. □

The results of [1] show that for quasi-free Hadamard states with relatively compact A and $\mathrm{dist}(A, B) > 0$, the relative entanglement entropy $E_R(\omega)$ between A and B is finite.

The expression for the operator α simplifies when the state is separating for $\mathfrak{A}_{B'}$ (e.g. when it is cyclic for $\mathfrak{A}_B$). In this case $\tilde{R}_A = P_A \leq P_{B'} = \tilde{R}_{B'}$ and

$$\alpha = \frac{1}{2} U_A \sqrt{P_{B'} - \Sigma_{B'}^2} \, U_A^* = j_A \alpha j_A.$$

Using the estimate $(P_A(P_{B'} - \Sigma_{B'}^2)^{\frac{1}{2}} P_A)^2 \leq P_A(P_{B'} - \Sigma_{B'}^2) P_A$ and taking quartic roots we find that $|(P_{B'} - \Sigma_{B'}^2)^{\frac{1}{4}} P_A| \leq |(P_{B'} - \Sigma_{B'}^2)^{\frac{1}{2}} P_A|^{\frac{1}{2}}$ and hence that

$$\sqrt{\alpha + j_A \alpha j_A} = U_A |(P_{B'} - \Sigma_{B'}^2)^{\frac{1}{4}} P_A| U_A^* \leq U_A |(P_{B'} - \Sigma_{B'}^2)^{\frac{1}{2}} P_A|^{\frac{1}{2}} U_A^*$$

and

$$E_R(\omega) \leq -4 \operatorname{Tr}_{\mathcal{K}(A)} \ln(1 - |(P_{B'} - \Sigma_{B'}^2)^{\frac{1}{2}} P_A|^{\frac{1}{2}}). \tag{4.34}$$

Let us now specialise to an ultra-static spacetime $\mathcal{M} = \mathbb{R} \times \mathcal{C}$ with a Cauchy surface $\mathcal{C}$ and metric $g = -dt^2 + h$ and to a massive $m > 0$ minimally coupled scalar field. Recall from Sect. 2.4.1 that in the ground state $\mathcal{K}$ can be identified with $W^{(\frac{1}{2})}(\mathcal{C}) \oplus W^{(-\frac{1}{2})}(\mathcal{C})$, where the Sobolev spaces are defined using the operator $C^{-1} := -\nabla^2 + m^2$, and that Σ takes the form given in Eq. (2.33). Analogously, we may identify $\mathcal{K}(A) = W^{(\frac{1}{2})}(A) \oplus W^{(-\frac{1}{2})}(A)$ and similarly for B and B', where $W^{\pm\frac{1}{2}}(A)$ is the closed subspace of $W^{\pm\frac{1}{2}}(\mathcal{C})$ generated by test-functions supported in A. We denote by $P_{A\pm}$ the orthogonal projections in $W^{\pm\frac{1}{2}}(\mathcal{C})$ onto $W^{\pm\frac{1}{2}}(A)$, so that $P_A = P_{A+} \oplus P_{A-}$, and similarly for B'.

Note that the ground state has the Reeh-Schlieder property, so we may use the estimate (4.34). Furthermore, using the facts that $\Sigma^2 = 1$ (which is true for any pure state) and $P_A \leq P_{B'}$ we find from the definition of $\Sigma_{B'}$ that

$$|(P_{B'} - \Sigma_{B'}^2)^{\frac{1}{2}} P_A|^2 = P_A \Sigma(1 - P_{B'}) \Sigma P_A = |(1 - P_{B'}) \Sigma P_A|^2$$

$$= \left| \begin{pmatrix} 0 & (1 - P_{B'+}) C^{\frac{1}{2}} P_{A-} \\ -(1 - P_{B'-}) C^{-\frac{1}{2}} P_{A+} & 0 \end{pmatrix} \right|$$

$$= \begin{pmatrix} |(1 - P_{B'-}) C^{-\frac{1}{2}} P_{A+}| & 0 \\ 0 & |(1 - P_{B'+}) C^{\frac{1}{2}} P_{A-}| \end{pmatrix}. \tag{4.35}$$

We may use the unitary operators $U_\pm : W^{\pm\frac{1}{2}}(\mathcal{C}) \to L^2(\mathcal{C})$ with $U_\pm f := C^{\mp\frac{1}{4}} f$ and $U_\pm^* f := C^{\pm\frac{1}{4}} f$ to rewrite this operator in terms of the L^2-inner product and the operator C. For this purpose we introduce the projections $Q_{A\pm} = U_\pm P_{A\pm} U_\pm$ in $L^2(\mathcal{C})$ and similarly for $Q_{B'\pm}$. Note that $Q_{A\pm}$ projects onto the subspace

$$\overline{\{C^{\mp\frac{1}{4}} f \mid f \in C_0^\infty(A)\}} \subset L^2(\mathcal{C}). \tag{4.36}$$

We then find from (4.34)[5]:

Proposition 7 *On an ultra-static spacetime $\mathcal{M} = \mathbb{R} \times \mathcal{C}$ with metric $g = -dt^2 + h$, the ground state ω_0 of a massive KG field fulfills*

$$E_R(\omega_0) \leq -4 \sum_\pm \operatorname{Tr} \ln(1 - |(1 - Q_{B'\mp}) Q_{A\pm}|^{\frac{1}{2}}). \tag{4.37}$$

---

[5]For ground states one may in fact omit the doubling procedure of [1] and directly apply the results of [13], which improves this estimate by a factor $\frac{1}{2}$.

Here the trace is in $L^2(\mathcal{C})$ and $Q_{A\pm}$ are the orthogonal projectors onto the closed subspaces (4.36), and similarly for $Q_{B'\pm}$.

Proof We use $Q_{A\pm} = U_\pm^* P_{A\pm} U_\pm$ and $U_\pm f := C^{\mp\frac{1}{4}} f$ in Eqs. (4.35) and (4.34), and the facts that $|UX| = |X|$ and $|XU| = U^*|X|U$ for all closed operators X and unitary operators U. Using the fact that C is bounded by m^{-2} one may show that $|(1 - Q_{B'\mp})Q_{A\pm}| < 1$ for both signs, so the right-hand side of (4.37) is well-defined. Note that the estimate clearly only depends on the geometry of $(\mathcal{C}, h)$ and the regions A and B.   □

We close this section proving our main estimate, which is an application of the previous proposition and some other techniques. This estimate shows that the entanglement entropy falls off exponentially with the distance between the regions A and B, in the following precise sense:

Theorem 10 *Consider an ultra-static spacetime $\mathcal{M} = \mathbb{R} \times \mathcal{C}$ with metric $g = -dt^2 + h$ and the ground state ω_0 of a minimally coupled free scalar field with mass $m > 0$. Let $A \subset \mathcal{C}$ be a bounded open region and for any $r > 0$ let $B_r := \{x \in \mathcal{C} \mid \mathrm{dist}(x, A) > r\}$. For any $R > 0$ and all $r \geq R$ the ground state ω_0 has an entanglement entropy between the regions A and B_r which satisfies*

$$E_R(\omega_0) \leq ce^{-\frac{1}{2}mr}$$

where $c > 0$ is independent of $r \geq R$.

Proof By the results of [1], the operators $|(1 - Q_{B'\mp})Q_{A\pm}|$ are compact, and their eigenvalues are in $[0, 1)$. It follows that the largest eigenvalue is less than 1, i.e. $\|(1 - Q_{B'\mp})Q_{A\pm}\| < 1$. For $t \in [0, a]$ with $a < 1$ we have the elementary estimate $-\ln(1 - t) < \frac{t}{1-a}$, which we may apply to $t = |(1 - Q_{B'\mp})Q_{A\pm}|^{\frac{1}{2}}$. It then follows from (4.37) that

$$E_R(\omega_0) \leq \sum_\pm \frac{4}{1 - \|(1 - Q_{B'\mp})Q_{A\pm}\|^{\frac{1}{2}}} \mathrm{Tr}\, |(1 - Q_{B'\mp})Q_{A\pm}|^{\frac{1}{2}}\,,$$

where the trace is over $L^2(\mathcal{C})$. As r increases, the projections $1 - Q_{B'\mp}$ decrease, so the prefactors $4(1 - \|(1 - Q_{B'\mp})Q_{A\pm}\|^{\frac{1}{2}})^{-1}$ attain their maximum at $r = R$. This value is still finite, because $R > 0$. We let c_0 denote the maximum of this value over the choices $+$ and $-$.

To estimate the traces we will use the fact that for all bounded linear operators X, Y we have $|XY| \leq \|X\| \cdot |Y|$ and $|XY| = V^*|Y^*X^*|V \leq \|Y^*\| \cdot V^*|X^*|V = \|Y\| \cdot V^*W|X|W^*V$, where V and W are the partial isometries appearing in the polar decompositions $XY = V|XY| = |(XY)^*|V$ and $X = W|X| = |X^*|W$.

Following [1] we choose a test-function χ_A with support in $B'_{\frac{1}{2}R}$ such that $\chi_A \equiv 1$ on A. We then choose test-functions χ_i, $i = 1, \ldots, 4$, such that $\chi_1 \equiv 1$ on $\mathrm{supp}(\chi_A)$ and $\chi_{i+1} \equiv 1$ on $\mathrm{supp}(\chi_i)$. We fix an $l \in \mathbb{N}$ such that $l > \frac{1}{2} + \frac{3}{4}\dim(\mathcal{C})$ and we introduce

$$X_i := \chi_i C^{(5-i)l} \chi_i C^{-(4-i)l}$$

for $i = 1, \ldots, 4$. The operators X_i are Hilbert-Schmidt (see [1] Theorem 4.2) and hence $X := X_1 X_2 X_3 X_4$ has $\operatorname{Tr} |X|^{\frac{1}{2}} < \infty$. Note that C^{-1} is a partial differential operator, so considering the supports of the χ_i we have

$$\chi_A C^{-4l-b} X = \chi_A C^{-b}$$

for any $b \in \mathbb{N}$. We fix b such that $b \pm \frac{1}{4} \geq 0$ and we use the fact that $Q_{A\pm} = C^{\mp \frac{1}{4}} \chi_A C^{\pm \frac{1}{4}} Q_{A\pm}$ in order to find

$$
\begin{aligned}
|(1 - Q_{B'_r \mp}) Q_{A\pm}| &= |(1 - Q_{B'_r \mp}) C^{\mp \frac{1}{4}} \chi_A C^{-b} C^{\pm \frac{1}{4}+b} Q_{A\pm}| \\
&= |(1 - Q_{B'_r \mp}) C^{\mp \frac{1}{4}} \chi_A C^{-4l-b} X C^{\pm \frac{1}{4}+b} Q_{A\pm}| \\
&\leq \|(1 - Q_{B'_r \mp}) C^{\mp \frac{1}{4}} \chi_A C^{-4l-b}\| \cdot \|C^{\pm \frac{1}{4}+b} Q_{A\pm}\| \cdot |X|
\end{aligned}
$$

and because $C^{\pm \frac{1}{4}+b}$ and $Q_{A\pm}$ are both bounded we have

$$\operatorname{Tr} |(1 - Q_{B'_r \mp}) Q_{A\pm}|^{\frac{1}{2}} \leq c_1 \|(1 - Q_{B'_r \mp}) C^{\mp \frac{1}{4}} \chi_A C^{-4l-b}\|^{\frac{1}{2}}$$

for some $c_1 > 0$ which is independent of r and of the sign $\pm$. Combining this estimate with the first paragraph of this proof we have

$$E_R(\omega_0) \leq c_0 c_1 \sum_{\pm} \|(1 - Q_{B'_r \mp}) C^{\mp \frac{1}{4}} \chi_A C^{-4l-b}\|^{\frac{1}{2}}. \tag{4.38}$$

For any test-function χ_B supported in B'_r and identically 1 on $B'_{\frac{1}{2}R}$ we have $(1 - Q_{B'_r \mp}) = (1 - Q_{B'_r \mp}) C^{\pm \frac{1}{4}} (1 - \chi_B) C^{\mp \frac{1}{4}}$ and hence

$$
\begin{aligned}
\|(1 - Q_{B'_r \mp}) C^{\mp \frac{1}{4}} \chi_A C^{-4l-b}\| &= \|(1 - Q_{B'_r \mp}) C^{\pm \frac{1}{4}} (1 - \chi_B) C^{\mp \frac{1}{2}} \chi_A C^{-4l-b}\| \\
&\leq \|C^{\pm \frac{1}{4}} (1 - \chi_B) C^{\mp \frac{1}{2}} \chi_A C^{-4l-b}\| \\
&= \|C^{-4l-b} \bar{\chi}_A C^{\mp \frac{1}{2}} (1 - \bar{\chi}_B) C^{\pm \frac{1}{4}}\|.
\end{aligned}
$$

Using Lemma 4.4 in [1] we may find $\eta_0, \ldots, \eta_{4l+b} \in C_0^\infty(B'_{\frac{1}{2}R})$ such that

$$\|C^{-4l-b} \bar{\chi}_A \psi\| \leq \left(\sum_{k=0}^{4l+b} \|\bar{\eta}_k C^{-k} \psi\|^2 \right)^{\frac{1}{2}} \leq \sum_{k=0}^{4l+b} \|\bar{\eta}_k C^{-k} \psi\|$$

for all ψ in the domain of C^{-4l-b}. Applying this result to $C^{\mp \frac{1}{2}} (1 - \bar{\chi}_B) C^{\pm \frac{1}{4}} f$ with test-functions f and using the fact that $\bar{\eta}_k C^{-k \mp \frac{1}{2}} (1 - \bar{\chi}_B) C^{\pm \frac{1}{4}}$ is bounded [1] Theorem 4.5) we find

$$\|(1 - Q_{B'_r\mp})C^{\mp\frac{1}{4}}\chi_A C^{-4l-b}\| \leq \sum_{k=0}^{4l+b} \|C^{\pm\frac{1}{4}}(1 - \chi_B)C^{-k\mp\frac{1}{2}}\eta_k\|. \qquad (4.39)$$

Note that the η_k are independent of r, as is the number of functions $4l + b + 1$.

For given $r \geq R$ we now choose a real-valued χ_B with support in B'_r such that $0 \leq \chi_B \leq 1$, $\chi_B \equiv 1$ on $B'_{r-\frac{1}{2}R}$ and such that $|\nabla\chi_B| \leq R$. For the $+$ sign it then follows immediately from $\|C\| \leq m^{-2}$ and Proposition 4.3 of [1] that

$$\sum_{k=0}^{4l+b} \|C^{\frac{1}{4}}(1 - \chi_B)C^{-k-\frac{1}{2}}\eta_k\| \leq c_2^2 e^{-mr}, \qquad (4.40)$$

because the supports of the η_k and $1 - \chi_B$ are separated by a distance $\geq r - R$. For the $-$ sign we note that

$$(1 - \chi_B)C^{-1}(1 - \chi_B) \leq C^{-1}\frac{(1 - \chi_B)^2}{2m^2}C^{-1} + \frac{(1 - \chi_B)^2 m^2}{2} + |\nabla\chi_B|^2,$$

where we estimated the term $-\nabla^2 \frac{(1-\chi_B)^2}{2m^2}\nabla^2 \leq 0$. Combining this estimate with the fact that $\|C^{-\frac{1}{4}}\psi\| \leq m^{-\frac{1}{2}}\|C^{-\frac{1}{2}}\psi\|$ for all ψ in the domain of $C^{-\frac{1}{2}}$, it then follows that

$$\|C^{-\frac{1}{4}}(1 - \chi_B)\psi\|^2 \leq \frac{1}{2m^2}\|(1 - \chi_B)C^{-1}\psi\|^2 + \frac{m^2}{2}\|(1 - \chi_B)\psi\|^2 + \||\nabla\chi|\psi\|^2$$

for all ψ in the domain of C^{-1}, where $|\nabla\chi| = \sqrt{\nabla^i\chi \cdot \nabla_i\chi}$ is a positive continuous function. We choose $\psi = C^{-k+\frac{1}{2}}\eta_k$ and applying Proposition 4.3 of [1] again, noting that its result also holds when the left function in the product is only continuous. We then find that

$$\sum_{k=0}^{4l+b} \|C^{-\frac{1}{4}}(1 - \chi_B)C^{-k+\frac{1}{2}}\eta_k\| \leq c_3^2 e^{-mr}, \qquad (4.41)$$

for some $c_3 \geq 0$, which is independent of r.

Putting together the estimates (4.39, 4.40, 4.41) (and taking square roots) yields

$$\|(1 - Q_{B'_r\mp})C^{\mp\frac{1}{4}}\chi_A C^{-4l-b}\|^{\frac{1}{2}} \leq \max\{c_2, c_3\}e^{-\frac{1}{2}mr}.$$

Inserting this into (4.38) and setting the constant in the theorem $c := 2c_0c_1 \max\{c_2, c_3\}$ completes the proof. $\qquad\qquad\qquad\qquad\qquad\qquad\qquad\qquad\qquad\qquad\qquad\square$

4.2.2 Free Dirac Fields

The BW-nuclearity condition is well-established for free massive Dirac fields in $d+1$ dimensional static spacetimes [5]. Thus, Theorem 6 applies and provides upper bounds on the entanglement entropy $E_R(\omega_0)$ of two diamonds of size r in Minkowski space separated by a distance R (or more generally, two bounded regions A and B in a static time-slice C separated by a distance R). One can again get better bounds for large R using techniques from modular theory in a similar way as for scalar fields [14].

Here we give an upper bound in the opposite regime when the distance, ε, between A and B goes to zero. Our reason for doing so is that this bound is qualitatively better than the general bound presented in Theorem 6—in fact it is of "area law" type. For simplicity, we focus on a Majorana field [cf. Sect. 2.4.2] on a static spacetime of the form $\mathscr{M} = \mathbb{R}^{1,1} \times \Sigma$, where (Σ, γ) is a compact $d - 1$-dimensional spin manifold, and where the metric on $\mathscr{M}$ is $g = -dt^2 + dx^2 + \gamma_{AB}(y)dy^A dy^B$. But our results presumably hold more generally when O_A is the interior of a black hole region in a spacetime with bifurcate Killing horizon[6] and O_B is a subset of the exterior, separated from the horizon by a corridor of diameter ε.

Our theorem is the following.

Theorem 11 *Let $B = \{x < 0, t = 0\}$ and $A = \{x > \varepsilon, t = 0\}$, and let ω_0 be the ground state of a free Majorana field on $\mathscr{M} = \mathbb{R}^{1,1} \times \Sigma$ of mass $m > 0$. Then for sufficiently small $\varepsilon > 0$, we have for every fixed $N \in \mathbb{N}, \delta > 0$ the upper bound*

$$E_R(\omega_0) \leq C |\ln(m'\varepsilon)| \sum_{j=d-1}^{-N} \varepsilon^{-j} \int_{\partial A} a_j(y), \qquad (4.42)$$

where $C > 0$ is a constant, $m' = 2m/[(1+\delta)(1+\delta^{-1})]^{\frac{1}{2}}$ and where a_j are the heat kernel expansion coefficients of the operator $(1+\delta)^{-\frac{1}{2}}(-\nabla_\Sigma^2 + \frac{1}{4}R_\Sigma)^{\frac{1}{2}}$ (see proof).

Remark 4 The first heat kernel coefficient a_{d-1} is constant (see e.g. [16]), so we get, to leading order for $\varepsilon \to 0$, the "area law",

$$E_R(\omega_0) \lesssim c_0 |\ln(m\varepsilon)| \frac{|\partial A|}{\varepsilon^{d-1}}. \qquad (4.43)$$

Proof Spacetime dimension $d + 1 = 2$: The essence of the proof is already seen in the case $d = 1$. The regions A and B and their causal completion O_A and O_B are in this case shown in Fig. 4.3 in Sect. 4.3 below (where R is replaced by ε here); they are left and right wedges. The construction of the algebra for the free Majorana field in $1 + 1$ dimensions was given above in Sect. 2.4.2, which we use here.

Let $|0\rangle$ be the vector representative of the vacuum state ω_0 in the GNS-Hilbert space $\mathcal{H}$. As usual, we consider $\Phi^A : \mathfrak{A}_A \to \mathcal{H}, a \mapsto \Delta_{B'}^{1/4} a|0\rangle$. We wish to apply

---

[6]See e.g. [15] for an explanation of this concept.

Theorems 3 and 4, which together with $\|\Psi^A\|_1 \le \|\Phi^A\|_1$ (Ψ^A given by (3.26)) give $E_R(\omega_0) \le \ln \|\Phi^A\|_1$.

So we need to estimate the nuclear 1-norm $\|\Phi^A\|_1$. The key point is that $\Delta_{B'}^{it}$ implements boosts on the Hilbert space $\mathcal{H}$, by the Bisognano-Wichmann theorem [17]. More precisely, let $U(\lambda)$ be the unitary implementer of the boost $\begin{pmatrix} \cosh\lambda & \sinh\lambda \\ \sinh\lambda & \cosh\lambda \end{pmatrix}$ given by Eq. (2.55) on fermonic Fock space. Then $\Delta_{B'}^{it} = U(-2\pi t)$. Now if $b' \in \mathfrak{A}_{B'}$, then since translations act geometrically on the algebras, $e^{-i\varepsilon P^1} b' e^{i\varepsilon P^1} = a \in \mathfrak{A}_A$ [here $e^{-i\varepsilon P^1}$ is the implementer (2.55) for a translation by $(0, \varepsilon)$ which maps the wedge O'_B to O_A by construction].

Therefore, the 1-norm of the map Φ^A is the same as the 1-norm of the map

$$\mathfrak{A}_{B'} \ni b' \mapsto \Xi^{B'}(b') = \Delta_{B'}^{\frac{1}{4}} e^{-i\varepsilon P^1} b'|0\rangle \in \mathcal{H}. \tag{4.44}$$

Due to the geometrical action of boosts, we have the identity

$$\Delta_{B'}^{it} e^{-i\varepsilon P^1} |\Psi\rangle = e^{-i\varepsilon \sinh(2\pi t) P^0 - i\varepsilon \cosh(2\pi t) P^1} \Delta_{B'}^{it} |\Psi\rangle. \tag{4.45}$$

for $|\Psi\rangle \in \mathrm{dom}\,\Delta'_B$. If we formally set $t = -i/4$, and put $|\Psi\rangle = b'|0\rangle$, we obtain

$$\Xi^{B'}(b') = e^{-\varepsilon P^0} \Delta_{B'}^{\frac{1}{4}} b'|0\rangle = e^{-\varepsilon P^0} U(i\pi/2) b'|0\rangle, \tag{4.46}$$

where $U(i\pi/2)$ is the representer of a boost with parameter λ analytically continued to $\lambda = i\pi/2$. A rigorous proof may be given using a similar argument as in Sect. 4.4; alternatively see [18].

Thus, at this stage, we have managed to show that $E_R(\omega_0) \le \ln \|\Xi^{B'}\|_1$, where $\Xi^{B'}$ is given by (4.46). We must thus understand this map. Since the operator $e^{-\varepsilon P^0} U(i\pi/2)$ is of "second quantized form" in a free field theory such as ours– it is given by Eq. 2.55 in general – it is plausible that that nuclear norm can be estimated using operators acting only on the 1-particle Hilbert space $\mathcal{H}_1 = L^2(\mathbb{R}, d\theta)$. That this is indeed the case can be be shown using results of [1], which are in turn based on results of [13, 18].

One first defines the *real* Hilbert space [recall $B' = (0, \infty)$]

$$H_{B'} = \{ e^{\theta - i\pi/4} \widetilde{k}_1(m \sinh\theta) + e^{-\theta + i\pi/4} \widetilde{k}_2(m \sinh\theta) \mid k_i \in C_0^\infty(B', \mathbb{R}) \} \subset L^2(\mathbb{R}, d\theta) \tag{4.47}$$

which is obtained by acting with $\psi(k)$ on $|0\rangle$ for *real*-valued test-spinors $k = (k_1, k_2)$ supported in $O_{B'}$, compare Eq. 2.48. On $\mathcal{H}_1$, we next introduce an anti-unitary "time reversal" operator $\widetilde{T}$. To this end, we first define, on $K = L^2(\mathbb{R}, dx; \mathbb{C}^2)$ the anti-linear operator $T : (k_1, k_2) \mapsto (\bar{k}_2, \bar{k}_1)$, which clearly satisfies $T^2 = 1$ (involution property). It follows that T commutes with the 1-particle hamiltonian h [see Eq. 62], and therefore also with the projector P onto the positive part of the spectrum of h. Therefore, T restricts to an operator on PK. Since, according to our discussion in Sect. 2.4.2, the 1-particle Hilbert space $\mathcal{H}_1 = L^2(\mathbb{R}, d\theta)$ is identified as VPK under the isometry (2.47), it follows that $VTV^* = \tilde{T}$ is an anti-unitary involution on $\mathcal{H}_1$. Next, we define the *complex* subspaces

$$H_{B'}^{\pm} = \mathbb{C} \cdot (1 \pm \tilde{T}) H_{B'} \subset \mathcal{H}_1. \tag{4.48}$$

Concretely, using the definition of V in Eq. (2.47) and of T, we have

$$
\begin{aligned}
H_{B'}^+ &= \{\Psi_k(\theta) = \cosh(\theta/2 - i\pi/4)\tilde{k}(m \sinh \theta) \mid \text{supp}\,(k) \subset B', k \in L^2(\mathbb{R})\}, \\
H_{B'}^- &= \{\Psi_k(\theta) = \sinh(\theta/2 - i\pi/4)\tilde{k}(m \sinh \theta) \mid \text{supp}\,(k) \subset B', k \in L^2(\mathbb{R})\}.
\end{aligned}
\tag{4.49}
$$

The wave functions in $H_A^{\pm}$ are in the domain of the operator $e^{-\varepsilon P^0} U(i\pi/2)$, and the action is in fact given by

$$[e^{-\varepsilon P^0} U(i\pi/2)\Psi_k](\theta) = e^{-m\varepsilon \cosh \theta}\Psi_k(\theta - i\tfrac{\pi}{2}) \equiv (X_1\Psi_k)(\theta), \tag{4.50}$$

where we note that the analytic continuation on the right side is indeed possible due to the support of k, and where we used Eq. (2.55) on fermonic Fock space with $n = 1$, $\lambda = i\pi/2$, and $a = (\varepsilon, 0)$. We omit the straightforward calculation. Theorem 3.11 of [1] now tells us that, if $E_{B'}^{\pm}$ are the projection operators onto the subspaces $H_{B'}^{\pm} \subset \mathcal{H}_1$, then we have the upper bound

$$\|\Xi^{B'}\|_1 \leq \exp\{2\|X_1 E_{B'}^+\|_1 + 2\|X_1 E_{B'}^-\|_1\}, \tag{4.51}$$

where the 1-norm on the right side is in $\mathcal{H}_1$. Similarly as in [18], one may use contour integration to rewrite the operator X_1 in the form

$$(X_1\Psi_k)(\theta) = \frac{-1}{2\pi i} e^{-m\varepsilon \cosh \theta} \int d\theta' \left\{ \frac{1}{\theta' - \theta + i\pi/2} + \frac{1}{\theta' + \theta - i\pi/2} \right\} \Psi_k(\theta') \tag{4.52}$$

when $\Psi_k \in H_{B'}^{\pm}$. For $\kappa, s \in \mathbb{R}, \kappa \neq 0, s > 0$, let us define an operator $T_{\kappa,s}$ on $L^2(\mathbb{R}, d\theta)$ given by the following kernel:

$$T_{\kappa,s}(\theta, \theta') = -\text{sign}(\kappa)\frac{e^{-\frac{1}{2}s \cosh \theta}}{2\pi i(\theta' - \theta + i\kappa/2)}. \tag{4.53}$$

In terms of this operator, we immediately get

$$\|X_1 E_{B'}^{\pm}\|_1 \leq \|T_{\pm\pi, 2m\varepsilon}\|_1. \tag{4.54}$$

The following lemma describes properties of the operator $T_{\kappa,s}$ needed to estimate the right side and also for later purposes:

Lemma 10 *1. In terms of the momentum operator $p = i\,d/d\theta$ on $L^2(\mathbb{R}, d\theta)$, we can write*

$$T_{\kappa,s} = e^{-\frac{1}{2}s \cosh \theta} \Theta(\kappa p) e^{-|\kappa p|/2} \tag{4.55}$$

where Θ is the Heaviside step function (characteristic function of the set $\mathbb{R}_+$).

2. The operator $A_{\kappa,s} = T_{+\kappa,s}T^*_{+\kappa,s} + T_{-\kappa,s}T^*_{-\kappa,s}$ has the integral kernel

$$A_{\kappa,s}(\theta,\theta') = \frac{|\kappa|}{\pi}\frac{e^{-\frac{1}{2}s\cosh\theta}e^{-\frac{1}{2}s\cosh\theta'}}{(\theta-\theta')^2 + \kappa^2}. \tag{4.56}$$

3.

$$\|T_{\kappa,s}\|_1 \lesssim \begin{cases} c_1 e^{-s/2} & for\, s \gg |\kappa| \\ c_2 \ln s & for\, s \ll |\kappa| \end{cases} \tag{4.57}$$

where c_i are numerical constants diverging no worse than $c_1 \lesssim |\ln\kappa|/|\kappa|^2$ respectively $c_2 \lesssim 1/|\kappa|^2$ for $|\kappa| \to 0$.

Combining now item (3) of Lemma 10 with (4.51), (4.54) gives, for $\varepsilon \ll m^{-1}$,

$$\ln\|\Phi^A\|_1 = \ln\|\Xi^{B'}\|_1 \le C|\ln(2m\varepsilon)|, \tag{4.58}$$

and since we have already noted that $E_R(\omega_0) \le \ln\|\Phi^A\|_1$, this proves the theorem when $d = 1$.

Spacetime dimension $d+1 > 2$: Let λ_j be the (real) eigenvalues of the elliptic operator $e_0 \cdot \sum_{A=2}^d e_A \cdot \nabla_{e_A}$ on the compact manifold Σ (enumerated without multiplicities), where e_A is a frame field for Σ. By decomposing a general spinor on $\mathcal{C} = \mathbb{R} \times \Sigma$ into the corresponding eigenmodes, one can easily show using the product structure $\mathcal{M} = \mathbb{R}^{1,1} \times \Sigma$ and Eq. 2.22 that the algebra $\mathfrak{A}_A$ is isomorphic to the tensor product $\otimes_j \mathfrak{A}_{A,j}$, where $\mathfrak{A}_{A,j}$ is isomorphic to the algebra for a Dirac field on $1 + 1$-dimensional Minkowski spacetime with mass $m_j = \sqrt{m^2 + \lambda_j^2}$. The analogous statement holds for B. If $\omega_{0,j}$ is the vacuum state for such a Dirac field on $\mathbb{R}^{1,1}$, property (e5) tells us that

$$E_R(\omega_0) \le \sum_j E_R(\omega_{0,j}). \tag{4.59}$$

Now let $\delta > 0$. Using our previous results in $d = 1$, and the trivial relation $m_j\sqrt{1+\delta} \ge \lambda_j\sqrt{1+\delta}^{-1} + m\sqrt{1+\delta^{-1}}^{-1}$ we also have

$$\begin{aligned} E_R(\omega_{0,j}) &\le 4\|T_{\pi,2m_j\varepsilon}\|_1 \\ &\le 4e^{-\varepsilon|\lambda_j|/(1+\delta)}\|T_{\pi,m'\varepsilon}\|_1 \le Ce^{-\varepsilon|\lambda_j|/(1+\delta)}\ln(\varepsilon m'), \end{aligned} \tag{4.60}$$

for $\varepsilon \ll m'$. Now we take the sum over j and use the relation

$$\sum_j e^{-t|\lambda_j|} = \mathrm{Tr}_{L^2(\Sigma,\$|_\Sigma)}\, e^{-t\sqrt{-\nabla_\Sigma^2 + \frac{1}{4}R_\Sigma}}. \tag{4.61}$$

The result then follows using well-known results on the heat kernel of elliptic pseudo-differential operators, see e.g. [16]. □

Proof of lemma 10: (1) and (2) follow by taking Fourier transforms. (3) Consider first $s \gg \kappa$, and take $\kappa > 0$ for definiteness (the other case is similar). We have using (1)

$$T_{\kappa,s} = e^{-\frac{1}{2}s\cosh\theta}e^{-\kappa p/2}\Theta(p)$$
$$= e^{-\frac{1}{2}(s-\kappa)\cosh\theta} \cdot e^{-\frac{1}{4}\kappa\cosh\theta}(p^2+1)^{-\frac{1}{2}} \cdot (p^2+1)^{\frac{1}{2}}e^{-\frac{1}{4}\kappa\cosh\theta}(p^2+1)^{-1}$$
$$\cdot (p^2+1)e^{-\kappa p/2}\Theta(p).$$

Now we apply the standard inequalities $\|XY\|_1 \le \|X\|_2\|Y\|_2$ and $\|XY\|_1 \le \|X\|$ $\|Y\|_1$ and use that $\|e^{-\frac{1}{2}(s-\kappa)\cosh\theta}\| \le e^{-\frac{1}{2}(s-\kappa)}$, $\|(p^2+1)e^{-\kappa p/2}\Theta(p)\| \le c_4\kappa^{-2}$ (for $|\kappa| < 2$). We get

$$\|T_{\kappa,s}\|_1 \le c_4\kappa^{-2}e^{-\frac{1}{2}(s-\kappa)}\|e^{-\frac{1}{4}\kappa\cosh\theta}(p^2+1)^{-\frac{1}{2}}\|_2\|(p^2+1)^{\frac{1}{2}}e^{-\frac{1}{4}\kappa\cosh\theta}(p^2+1)^{-1}\|_2. \tag{4.62}$$

Using $\frac{1}{4\pi}e^{-|\theta|} = \frac{1}{2\pi}\int_\mathbb{R} dp(1+p^2)^{-1}e^{ip\theta}$ for the integral kernel of $(1+p^2)^{-1}$, we have

$$\|e^{-\frac{1}{4}\kappa\cosh\theta}(p^2+1)^{-\frac{1}{2}}\|_2^2 = \frac{1}{4\pi}\int d\theta d\theta' e^{-\frac{1}{4}\kappa\cosh\theta - \frac{1}{4}\kappa\cosh\theta' - |\theta-\theta'|} \le c_5(1+|\ln\kappa|) \tag{4.63}$$

for some c_5 independent of κ. The norm $\|(p^2+1)^{\frac{1}{2}}e^{-\frac{1}{4}\kappa\cosh\theta}(p^2+1)^{-1}\|_2$ can be estimated in a similar way noting that $p^2 + 1$ is just a differential operator. This proves the lemma in the case $s \gg \kappa$. The other case is treated similarly. □

4.3   Upper Bounds for Integrable Models

Here we apply the general upper bounds presented in Theorems 3 and 4 to the case of certain integrable models in $1 + 1$ dimensions with factorizing S-matrix described in Sect. 2.4.3.

We denote points in $1 + 1$ dimensional Minkowski space by (t, x). For A we choose $A = \{t = 0, x > R\}$ and for B be choose $B = \{t = 0, x < 0\}$. The regions O_A and O_B are obviously left and right wedges translated by $R > 0$, i.e. $O_A = \{x - R > |t|\}$ and $O_B = \{-x > |t|\}$. The distance between A and B is R, see Fig. 4.3.

As usual, we set $\mathfrak{A}_A = \pi_0(\mathfrak{A}(O_A))''$, and similarly for $\mathfrak{A}_B$, where π_0 is the GNS-representation of the vacuum state, ω_0, of the model. For definiteness, we consider integrable models defined via a factorizing 2-body S-matrix of the form

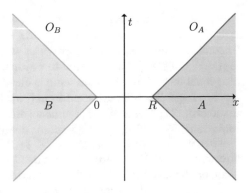

Fig. 4.3 The wedge regions O_A, O_B

$$S_2(\theta) = \prod_{k=1}^{2N+1} \frac{\sinh \theta - i \sin b_k}{\sinh \theta + i \sin b_k}, \tag{4.64}$$

where each $0 < b_i < \pi/2$. This S_2-matrix has the required properties (s1)–(s4) and therefore defines a net of v. Neumann algebras as described in Sect. 2.4.3. Our theorem is:

Theorem 12 *Let ω_0 be the vacuum state of the integrable quantum field theory of mass m defined by S_2, and let $A = \{t = 0, x > R\}$ and $B = \{t = 0, x < 0\}$. For $mR \gg 1/(\kappa\delta)$, we have for any κ with $\min\{b_i\} > \kappa > 0$ and any $\delta > 0$*

$$E_R(\omega_0) \lesssim \frac{4ec}{\kappa\sqrt{\pi mR}} e^{-mR(1-\delta)} \tag{4.65}$$

where $c = \sup\{|S_2(\zeta)| : -\kappa < \Im\zeta < \pi + \kappa\}^{1/2}$ is a constant depending on S_2, κ.

Remark 5 (1) For $mR \ll 1$, our estimations do not produce a bound so far. Looking at the proof, it is clear that an improved estimation of the quantity $\mathrm{Tr} \wedge^n A_{\kappa,s}$ appearing below in Eq. (4.77) for $\kappa \ll 1$ is necessary. We conjecture that this would lead for $mR \ll 1$ to a bound of the type

$$E_R(\omega_0) \lesssim C |\ln mR|^\alpha \tag{4.66}$$

for some constants α, C depending on S_2. One may also guess that E_R is perhaps asymptotic to the v. Neumann entropy for A with a UV-cutoff $\sim R$. This v. Neumann entropy has been computed using the replica trick (and also approximations) in [19]. From that work we thus perhaps expect the sharp upper bound to be of the form $\sim (c_{\mathrm{eff}}/6) \ln(mR)$, where c_{eff} is an effective UV central charge defined in that work. (2) Our upper bounds also hold for any pair of diamond regions $O_{A/B}$ that are space-like separated by a Lorentz invariant distance $R = \mathrm{dist}(O_A, O_B)$, due to the monotonicity property of the relative entanglement entropy, (e4), and the Lorentz-invariance of ω_0.

Proof Let $|0\rangle$ be the vector representative of the vacuum state ω_0 in the GNS-Hilbert space $\mathcal{H}$. As usual, we consider $\Phi^A : \mathfrak{A}_A \to \mathcal{H}, a \mapsto \Delta_{B'}^{1/4} a|0\rangle$. We wish to apply Theorems 3 and 4, which together with $\|\Psi^A\|_1 \leq \|\Phi^A\|_1$ (Ψ^A given by (3.26)) give $E_R(\omega_0) \leq \ln \|\Phi^A\|_1$. So we need to estimate the nuclear 1-norm $\|\Phi^A\|_1$, which is also equal to $\|\Phi^B\|_1$, since the setup is clearly symmetric in A, B. These estimates can to a large extent be extracted from [20], which in turn improve and correct estimations in [21]. Some of our arguments differ somewhat and may lead to important improvements of [20], so we give some details.

First, we observe that $\|\Phi^A\|_1 = \|\Xi^{B'}\|_1$, where $\Xi^{B'} : \mathfrak{A}_{B'} \to \mathcal{H}$ is the map defined already above in Eq. (4.46). The argument is precisely the same as there and relies again on the fact that $\Delta_{B'}^{it}$ are the generators of boosts of the wedge $O_{B'}$.

Let $\Xi_n = P_n \Xi^{B'}$, where $P_n : \mathcal{H} \to \mathcal{H}_n$ is the orthogonal projector onto the n-particle subspace of the S_2-symmetric Fock space, see Sect. 2.4.3. Using Eq. (4.46) and using the action of boosts and translations on wave functions in $\mathcal{H}_n$ described in Sect. 2.4.3, Eq. 2.55, we immediately get

$$\Xi_n(b') = \prod_{j=1}^{n} e^{-mR\cosh\theta_j} \cdot (P_n b'|0\rangle)(\theta_1 - i\tfrac{\pi}{2}, \ldots, \theta_n - i\tfrac{\pi}{2}). \tag{4.67}$$

Following [20, 21], we now consider the decomposition

$$\Xi_n = E_n \circ X_n \circ \Upsilon_n. \tag{4.68}$$

(a) $\Upsilon_n : \mathfrak{A}_{B'} \to H^2(\mathbb{R}^n + iC_n)$ is the map into the Hardy space defined by ($\delta > 0$ small)

$$\Upsilon_n : b' \mapsto \prod_{j=1}^{n} e^{-i\delta mR\sinh\zeta_j} \cdot (P_n b'|0\rangle)(\zeta_1, \ldots, \zeta_n). \tag{4.69}$$

Here, $(\zeta_1, \ldots, \zeta_n) \in \mathbb{R}^n + iC_n \subset \mathbb{C}^n$ is an n-tuple of complex numbers, C_n is a suitable open polyhedron in $\mathbb{R}^n$, and the Hardy space is defined to be the Banach space of those holomorphic functions $h(\zeta_1, \ldots, \zeta_n)$ having finite Hardy norm

$$\|h\|_{H^2}^2 = \sup_{(\lambda_1,\ldots,\lambda_n)\in C_n} \int_{\mathbb{R}^n} d^n\theta \, |h(\theta_1 + i\lambda_1, \ldots, \theta_n + i\lambda_n)|^2. \tag{4.70}$$

It can be shown using $b' \in \mathfrak{A}_{B'}$ that the analytic continuation of the n-particle wave function $(P_n b'|0\rangle)(\theta_1, \ldots, \theta_n)$ to $\mathbb{R}^n + iC_n$ is possible e.g. for the choice $C_n = (-\tfrac{\pi}{2}, \ldots, -\tfrac{\pi}{2}) + (-\tfrac{\kappa}{2n}, \tfrac{\kappa}{2n})^{\times n}$, with κ as in the hypothesis of the theorem.

(b) $X_n : H^2(\mathbb{R}^n + iC_n) \to L^2(\mathbb{R}^n)$ is the map defined by

$$X_n : h(\zeta_1, \ldots, \zeta_n) \mapsto \prod_{j=1}^{n} e^{-(1-\delta)mR\cosh\theta_j} \cdot h(\theta_1 - i\tfrac{\pi}{2}, \ldots, \theta_n - i\tfrac{\pi}{2}), \tag{4.71}$$

and E_n is the projector onto the S_2-symmetric wave functions described in Sect. 2.4.3. Next, we need bounds on the norms of $E_n \circ X_n$ respectively Υ_n:

(a) Combining Proposition 3.8 and Lemma 5.1 of [20], and (1), (2) of Lemma 10, one gets the following upper bound. There is a constant c depending only on κ and S_2 such that

$$\|E_n \circ X_n\|_1 \leq c^n \sum_{\sigma_1,\ldots,\sigma_n=\pm 1} \mathrm{Tr} \left(\bigwedge_{j=1}^{n} R_{\sigma_j,\kappa/2n,(1-\delta)mR} \right) = c^n \, \mathrm{Tr}(\wedge^n A_{\kappa/2n,(1-\delta)mR}),$$

(4.72)

where $R_{\pm,\kappa,s} = T_{\pm\kappa,s} T^*_{\pm\kappa,s}$ with $T_{\kappa,s}$ the operator in Lemma 10, and where $A_{\kappa,s} = R_{+,\kappa,s} + R_{-,\kappa,s}$ is equal to the operator defined in item (2) of Lemma 10. The n-th exterior power $\wedge^n A$ of an operator A on $L^2(\mathbb{R}, d\theta)$ by definition means the restriction of $\otimes^n A$ to the subspace of totally *anti-symmetric* (*not* S_2-*symmetric!*) wave functions in $L^2(\mathbb{R}^n, d^n\theta)$. It is also shown in Lemma 5.7 of [21] that the constant c may be chosen as $c = \|S_2\|_\kappa^{1/2}$, using the shorthand $\|S_2\|_\kappa = \sup_\kappa\{|S_2(\zeta)|\}$, with the supremum taken in the strip $-\kappa < \Im\zeta < \pi + \kappa$.

(b) Furthermore, it is shown in Proposition 4.5 of [20] that

$$\|\Upsilon_n\| \leq \max \left\{ 1, \|S_2\|_\kappa^n \left(\frac{2}{\pi\kappa} \int_0^\infty d\theta \, e^{-mR\delta \sin\kappa \cosh\theta} \right)^{\frac{1}{2}} \right\},$$

(4.73)

These upper bounds will now be applied to the right side of

$$\|\Phi^A\|_1 = \|\Xi^{B'}\|_1 \leq \sum_{n=0}^\infty \|\Xi_n\|_1 \leq \sum_{n=0}^\infty \|E_n \circ X_n\|_1 \|\Upsilon_n\|.$$

(4.74)

To get a better handle on the right side of the bound (4.72), we next use the following two lemmas:

Lemma 11 *Let A be a positive trace-class operator on $L^2(\mathbb{R}, d\theta)$ with smooth integral kernel $A(\theta, \theta')$. Then there holds*

$$\mathrm{Tr}(\wedge^n A) = \frac{1}{n!} \int_{\mathbb{R}^n} d^n\theta \, \det[A(\theta_i, \theta_j)]_{1 \leq i,j \leq n}.$$

(4.75)

Proof Obvious generalization of well-known formula in statistical mechanics, see e.g. Sect. 7.2 in [22]. □

Lemma 12 *Let T be a complex, positive definite $n \times n$ matrix. Then*

$$\det T \leq \prod_{i=1}^{n} T_{ii}.$$

(4.76)

Proof Well-known. Follows e.g. from "Gram's" or "Hadamard's inequality". □

We apply these two lemmas to the operator $A = A_{\kappa,s}$ defined in item (2) of Lemma 10. Applying first Lemma 11 gives

$$\mathrm{Tr}\wedge^n A_{\kappa,s} = \frac{\kappa^n}{\pi^n n!} \int_{\mathbb{R}^n} \mathrm{d}^n\theta \, \det[(\theta_i - \theta_j)^2 + \kappa^2]^{-1} \prod_{j=1}^{n} e^{-s\cosh\theta_j}. \qquad (4.77)$$

The integral kernel $[(\theta - \theta')^2 + \kappa^2]^{-1}$ is positive definite (i.e. gives a positive operator) because its Fourier transform is a positive constant times the function $e^{-\kappa|p|} > 0$. It follows from standard characterizations of positive definite kernels that $T_{ij} = [(\theta_i - \theta_j)^2 + \kappa^2]^{-1}$ is a positive $n \times n$ matrix for any choice of $\{\theta_j\}$. Therefore Lemma 12 gives $|\det T| \leq \kappa^{-2n}$. This immediately results in

$$\mathrm{Tr}\wedge^n A_{\kappa,s} \leq \frac{1}{\pi^n \kappa^n n!} \int_{\mathbb{R}^n} \mathrm{d}^n\theta \prod_{j=1}^{n} e^{-s\cosh\theta_j} = \frac{1}{n!} \left(\frac{2}{\kappa\pi} K_0(s)\right)^n, \qquad (4.78)$$

using the representation $\int_0^\infty \mathrm{d}\theta \, e^{-x\cosh\theta} = K_0(x)$, $x > 0$ of the Bessel function K_0. Hence, (4.72) gives us:

$$\|E_n \circ X_n\|_1 \leq \frac{1}{n!}\left(\frac{4nc}{\kappa\pi} K_0[(1-\delta)mR]\right)^n \leq \left(\frac{4ec}{\kappa\pi} K_0[(1-\delta)mR]\right)^n \qquad (4.79)$$

for $n \geq 1$ (using Stirling's approximation $n^n \leq e^n n!$). From (4.74), (4.73), we then get the bound

$$\|\Xi^{B'}\|_1 \leq \sum_{n=0}^{\infty}\{c_1 K_0[(1-\delta)mR]\}^n \max\left\{1, c_2^n[K_0(mR\delta\sin\kappa)]^{\frac{1}{2}}\right\} \qquad (4.80)$$

where $c_1 = \frac{4ec}{\kappa\pi}$, $c = \|S_2\|_\kappa^{1/2}$, c_2 are constants depending on S_2 and κ which will diverge when $\kappa \to 0$. For the Bessel function K_0 it is well-known that

$$K_0(x) \sim \begin{cases} -\ln x & \text{for } x \to 0^+ \\ \sqrt{\pi/x}\,e^{-x} & \text{for } x \to \infty. \end{cases} \qquad (4.81)$$

We can now discuss the asymptotic behavior of $\|\Xi^{B'}\|_1$. For $0 < mR \ll 1$, the right side of (4.80) does not converge, so we are unable to obtain a bound on $\|\Xi^{B'}\|_1$ in that case. For $mR \gg 1/\kappa\delta$, instead we get convergence, and in fact,

$$(E_R(\omega_0) \leq) \ln\|\Xi^{B'}\|_1 \lesssim \frac{4e}{\kappa}\sqrt{\frac{\|S_2\|_\kappa}{\pi mR}} e^{-mR(1-\delta)}, \qquad (4.82)$$

as claimed. $\square$

4.4 Upper Bounds for Conformal QFTs in $d + 1$ Dimensions

Here we apply our methods to derive a general bound on $E_R(\omega_0)$ for the vacuum state of a conformal field theory (CFT) in $d + 1$ dimensions, where $d > 1$ (the case when $d = 1$ is somewhat special and is treated in Sect. 4.6). In the case of conformal quantum field theories, the axioms (a3), (a4) are suitably extended to the conformal group, $G = SO_+(d + 1, 2)/\mathbb{Z}_2$ of $d + 1$-dimensional Minkowski spacetime. The action of G on points $x \in \mathbb{R}^{d,1}$ can be efficiently described in the well-known "embedding formalism", wherein one considers first the action of G on the cone $C = \{\xi \in \mathbb{R}^{d+3} \mid \tilde{\eta}(\xi, \xi) = 0\}$, where $\tilde{\eta} = diag(-1, 1, \ldots, 1, -1)$ is the metric of signature $(-2, d + 1)$ on $\mathbb{R}^{d+1,2}$. The projective cone $C/\mathbb{R}_\times = \overline{\mathbb{R}^{d,1}}$ is the Dirac-Weyl compactification (see e.g. [23]) of Minkowski space, and on this compactification, the action of G is defined *globally*. Uncompactified Minkowski space can be identified with the subset of points for which $x^\mu = \xi^\mu/(\xi^{d+1} + \xi^{d+2})$, $\mu = 0, 1, \ldots, d$ is finite. The transformation $\xi \mapsto g\xi$ induces a *local* action of G on $\mathbb{R}^{d,1}$ via this identification, which we write $x \mapsto g \cdot x$, where local means that it is not defined for all pairs g, x. The geometric significance of the various subgroups of G can be described as follows:

(i) The subgroup leaving ξ^{d+1}, ξ^{d+2} fixed given by $\xi'^\mu = \Lambda^\mu{}_\nu \xi^\nu$, where Λ is a proper orthochronous Lorentz transformation.

(ii) The $d + 1$-parametric subgroup of transformations $\xi'^\mu = \xi^\mu + \xi^+ a^\mu, \xi'^- = \xi^-, \xi'^+ = \xi^+ + 2\xi_\mu a^\mu + \xi^- a_\mu a^\mu$ corresponding to translations by a^μ. Here $\xi^\pm = \xi^{d+1} \pm \xi^{d+2}$.

(iii) The $d + 1$-parametric subgroup of transformations $\xi'^\mu = \xi^\mu - \xi^- c^\mu, \xi'^+ = \xi^+, \xi'^- = \xi^- - 2\xi_\mu c^\mu + \xi^+ a_\mu c^\mu$ corresponding to special conformal transformation with parameters c^μ.

(iv) The dilations by $\lambda > 0$ correspond to $\xi'^\mu = \xi^\mu, \xi^\pm = \lambda^{\mp 1} \xi^\pm$.

Since, by contrast with the action of the Poincaré group, conformal transformations cannot be globally defined for all pairs g, x, when stating the axioms of covariance in conformal field theory, one can at best require covariance only for orbits of points which do not pass trough infinity[7]:

(a3′) (Conformal invariance) For any finitely extended region $U \subset \mathbb{R}^{d,1}$ there exists a neighborhood $N \subset G$ of the identity such that for any $g \in N$, one has an algebraic isomorphism α_g respecting the net structure in the sense that $\alpha_g \mathfrak{A}(O) = \mathfrak{A}(g \cdot O)$ for all causal diamonds $O \subset U$. For $g, g' \in N$ such that also $gg' \in N$, there holds $\alpha_g \alpha_{g'} = \alpha_{gg'}$.

The vacuum axiom becomes:

(a4′) (Vacuum) There is a unique state ω_0 on $\mathfrak{A}$ such that $\omega_0(\alpha_g(a)) = \omega_0(a)$ whenever $\alpha_g(a)$ is defined. On its GNS-representation $(\pi_0, \mathcal{H}_0, |0\rangle)$, α_g is imple-

---

[7]Alternatively, one may pass to a net on the Dirac-Weyl compactification, as described in [23].

mented by a strongly continuous projective positive energy[8] representation U of the covering group $\widetilde{G}$ in the sense that

$$U(g)\pi_0(a)|0\rangle = \pi_0(\alpha_g(a))|0\rangle, \tag{4.83}$$

whenever $\alpha_g(a)$ is defined.

Denoting the generators of the Lie algebra $\mathfrak{so}(d + 1, 2)$ by M_{AB}, $A, B = 0, \ldots,$ $d + 2$ with relations

$$[M_{AB}, M_{CD}] = 2(\tilde{\eta}_{A[C} M_{D]B} - \tilde{\eta}_{B[C} M_{D]C}), \tag{4.84}$$

we define the following self-adjoint generators on $\mathcal{H}_0$:

$$P^0 = \frac{1}{i}\frac{d}{dt}U(\exp t(M_{0(d+2)} + M_{0(d+1)}))\Big|_{t=0},$$
$$K^0 = \frac{1}{i}\frac{d}{dt}U(\exp t(M_{0(d+2)} - M_{0(d+1)}))\Big|_{t=0}, \tag{4.85}$$
$$D = \frac{1}{i}\frac{d}{dt}U(\exp t M_{(d+1)(d+2)})\Big|_{t=0},$$

which are the generators of time-translations (ii), special conformal transformations (iii) in the time-direction, and dilations (iv), respectively.

We can now state the first result of this subsection. Let $R > r > 0$ and let A be a ball of radius r, B be the complement a ball of radius R centered at the origin in a time slice $\mathbb{R}^d$ of $d + 1$-dimensional Minkowski spacetime, see Fig. 1.2.

As usual, let O_A, O_B be the corresponding domains of dependence, and $\mathfrak{A}_A = \pi_0(\mathfrak{A}(O_A))''$ and $\mathfrak{A}_B = \pi_0(\mathfrak{A}(O_B))''$ the corresponding v. Neumann algebras of observables acting on the vacuum Hilbert space $\mathcal{H}_0$.

Theorem 13 *Let ω_0 be the vacuum state. We have*

$$E_R(\omega_0) \leq \ln \operatorname{Tr}\left(\frac{r}{R}\right)^{\frac{1}{2}(P^0 + K^0)}. \tag{4.86}$$

Proof Since the theory is invariant under dilations, it is clearly sufficient to prove the theorem in the special case $R = 1 > r > 0$. We would like to apply Theorems 3 and 4. We note again that $\pi(\mathfrak{A}(O_{B'}))'' \subset \mathfrak{A}'_B$, it follows $\Delta \leq \Delta_{B'}$, where Δ is the modular operator for $\mathfrak{A}'_B$ considered before in condition (3.26), so that $\|\Psi^A\|_1 \leq \|\Phi^A\|_1$. This shows that $E_R(\omega_0) \leq \ln \|\Phi^A\|_1$, where $\Phi^A : \mathfrak{A}_A \to \mathcal{H}_0$ is defined as usual by $a \mapsto \Delta_{B'}^{\frac{1}{4}} a |0\rangle$. Here $B' = \mathbb{R}^d \setminus \overline{B}$, as usual. So we need to estimate $\|\Phi^A\|_1$. Inspired by [25], we define an operator T by

---

[8]This means as usual that P^0 has non-negative spectrum. It can be shown that this implies that also the "conformal Hamiltonian" $\frac{1}{2}(P^0 + K^0)$ appearing below has non-negative spectrum [24]. The relation $e^{isD}(P^0 + K^0)e^{-isD} = e^{-s}P^0 + e^{+s}K^0$ then implies the same for K^0 (letting $s \to \infty$).

$$T = \Delta_{B'}^{\frac{1}{4}} \Delta_A^{-\frac{1}{4}}. \tag{4.87}$$

We would like to show that

$$T = r^{\frac{1}{2}(P^0 + K^0)} r^{iD} \tag{4.88}$$

on a dense core of vectors in $\mathcal{H}_0$. Let us assume this has been done. Then, since D is self-adjoint, we clearly have

$$\sqrt{TT^*} = r^{\frac{1}{2}(P^0 + K^0)}. \tag{4.89}$$

It follows that

$$\|T\|_1 = \|\sqrt{TT^*}\|_1 = \mathrm{Tr}\ r^{\frac{1}{2}(P^0 + K^0)}. \tag{4.90}$$

Now, as we will argue momentarily, the map $\Xi^A : \mathfrak{A}_A \to \mathcal{H}_0$ given by $\Xi^A(a) = \Delta_A^{\frac{1}{4}} a |0\rangle$ has norm $\|\Xi^A\| \leq 1$, so the above operator identity for T gives us,

$$\|\Phi^A\|_1 = \|T\Xi^A\|_1 \leq \|T\|_1 \cdot \|\Xi^A\| \leq \mathrm{Tr}\ r^{\frac{1}{2}(P^0 + K^0)}, \tag{4.91}$$

and therefore $E_R(\omega_0) \leq \ln \mathrm{Tr}\ r^{\frac{1}{2}(P^0 + K^0)}$, thereby showing the theorem if we can demonstrate $\|\Xi^A\| \leq 1$. This can be demonstrated by the following standard argument. Consider the function appearing in the KMS-condition, $f_a(z) = \langle 0|a^* \Delta_A^{-iz} a|0\rangle$. By Proposition 1, it is analytic on the strip $0 < \Im(z) < 1$ and continuous at the boundary of the strip. For the boundary value at $\Im(z) = 0$ one finds $|f_a(t)| \leq \|a|0\rangle\|^2 \leq \|a\|^2$. For the boundary value at $\Im(z) = 1$, one can use the KMS-condition to find $|f_a(t + i)| = |f_{a^*}(-t)| \leq \|a\|^2$. We can now apply the three line-theorem and conclude that $|f_a(z)| \leq \|a\|^2$ also in the interior of the strip. The value $z = i/2$ gives the result because $f_a(i/2) = \|\Xi^A(a)\|^2$, so indeed $\|\Xi^A(a)\| \leq \|a\|$.

We still need to demonstrate the operator identity (4.88). For this, we first formulate a lemma:

Lemma 13 *For $s, t \in \mathbb{R}$, we have the operator identity*

$$\Delta_{B'}^{it} e^{isD} \Delta_{B'}^{-it} = \exp\left[-\frac{is}{2} \sinh(2\pi t)(P^0 + K^0) + is \cosh(2\pi t) D \right] \tag{4.92}$$

on $\mathcal{H}_0$.

Proof To prove this formula, we use that the modular operators of double cones act in a geometrical way in conformal quantum field theories according to the Hislop-Longo theorem [23, 26].[9] The precise result is as follows. Let $L(t)$ be the 1-parameter family of conformal transformations defined by

---

[9] The proof was given in [26] for a free massless scalar field. It was subsequently shown by Brunetti, Guido and Longo [23] that the theorem generalizes to theories that fit into our axiomatic setting for CFTs.

$$L(t) = \exp(-2\pi t M_{0(d+1)}) = \begin{pmatrix} \cosh(2\pi t) & 0 & \ldots & -\sinh(2\pi t) & 0 \\ 0 & 1 & \ldots & 0 & 0 \\ \vdots & \vdots & & \vdots & \vdots \\ -\sinh(2\pi t) & 0 & \ldots & \cosh(2\pi t) & 0 \\ 0 & 0 & \ldots & 0 & 1 \end{pmatrix}. \qquad (4.93)$$

Then the theorem [23, 26] is that $\Delta_{B'}^{it} = U(L(t))$. Recalling that the generator of dilations in the Lie algebra $\mathfrak{so}(d+1,2)$ is given by $M_{(d+1)(d+2)}$, we compute in G

$$\mathrm{Ad}(L(t))M_{(d+1)(d+2)} = -\sinh(2\pi t)M_{0(d+2)} + \cosh(2\pi t)M_{(d+1)(d+2)}. \qquad (4.94)$$

Applying the unitary representation U, and recalling the defining relations (4.85) for D, P^0, K^0 this immediately gives the statement of the lemma.   □

Next, since D generates dilations, and since region A is obtained from region B' by shrinking B' by r, it is geometrically clear (and can easily be proven) that

$$\Delta_A = e^{-i(\ln r)D} \Delta_{B'} e^{i(\ln r)D}, \qquad (4.95)$$

so taking $s = -\ln r$ in the lemma and multiplying the formula in the lemma by $e^{i(\ln r)D}$ from the right, we get

$$\Delta_{B'}^{it} \Delta_A^{-it} = \exp\left[+\frac{i}{2} \ln(r) \sinh(2\pi t)(P^0 + K^0) - i\ln(r)\cosh(2\pi t)D \right] e^{i(\ln r)D} \qquad (4.96)$$

in $\mathcal{H}_0$. The desired formula (4.88) now formally follows by setting $t = -i/4$.

Justifying this last step occupies the remainder of this proof. Generalizing (4.87), let us set $T_s(z) = \Delta_{B'}^z \Delta_A^{-z}$, where z is in the strip $0 \le \Re(z) \le \frac{1}{2}$. We include a subscript "s" here to emphasize the dependence on the regions A, B, since their relative positions are fixed by $r = e^{-s} < 1$. Since $\mathfrak{A}_{B'} \supset \mathfrak{A}_A$, it follows that (Lemma 2.9 of [1], which uses a generalization of the Heinz-Löwner theorem [27] to unbounded operators) $\Delta_{B'}^\alpha \le \Delta_A^\alpha$ for $0 \le \alpha \le 1$. Therefore, $T(\alpha/2)^* T(\alpha/2) \le 1$ for this range of α. It then also follows that $T_s(z)$ is holomorphic in the strip $0 < \Re(z) < \frac{1}{2}$ and continuous on its closure, and that $\|T_s(z)\| = \|\Delta_{B'}^{i\Im z} T_s(\Re z) \Delta_A^{-i\Im z}\| \le 1$ for $0 \le \Re(z) \le \frac{1}{2}$.

Now let

$$f_1(z) = \Delta_{B'}^z e^{isD} \Delta_{B'}^{-z} |\chi\rangle = T_s(z) e^{isD} |\chi\rangle,$$
$$f_2(z) = \exp\left[\frac{is}{2} \sinh(2\pi iz)(P^0 + K^0) + is\cosh(2\pi iz)D \right] |\chi\rangle, \qquad (4.97)$$

defined first for z such that $\Re(z) = 0$. By what we have just said about $T_s(z)$, the function $f_1(z)$ has an analytic continuation to the strip $0 < \Re(z) < \frac{1}{4}$ that is continuous on its closure. f_2 has an analytic continuation to an open complex neighborhood of the form $\{z \mid |z| < \frac{1}{2}\}$ in the complex plane for a dense set of ("smooth") vectors

$|\chi\rangle \in \mathcal{H}_0$ [28], provided $|s|$ is sufficiently small. By Lemma 13, f_1 and f_2 agree on the imaginary axis. It follows by the edge of the wedge theorem (see Appendix A.1) that f_1 and f_2 coincide in the open neighborhood of $\{z \in \mathbb{C} \mid 0 < \Re(z) < \frac{1}{2}, |z| < \frac{1}{2}\}$ and by continuity on its closure. Thus we may take $z = \frac{1}{4}$, and it follows

$$\Delta_{B'}^{\frac{1}{4}} e^{isD} \Delta_{B'}^{-\frac{1}{4}} |\chi\rangle = \exp\left[-\frac{s}{2}(P^0 + K^0)\right]|\chi\rangle \tag{4.98}$$

for sufficiently small $|s|$ and all smooth vectors $|\chi\rangle$. Since the operators are bounded for $s > 0$, the formula in fact holds for all $|\chi\rangle \in \mathcal{H}_0$ when $s > 0$ is sufficiently small. Furthermore, both sides define 1-parameter semi-groups in s, so the identity holds also for all $s > 0$. The desired operator identity (4.88) now follows setting $s = -\ln(r) > 0$ in view of (4.95). $\square$

Our next aim is to relate the "partition function" on the right side of our bound (4.86) to the "spectrum of operator dimensions" in the given CFT. To state our result, we need to describe our conformal field theory in terms of quantum fields $\mathcal{O}(x)$, which are unbounded operator-valued distributions. Given an algebraic quantum field theory described by a net of observables algebras satisfying (a1)–(a5), [29] have shown how to define a set of linearly independent operator valued distributions

$$\Phi = \left\{ \mathcal{O} : f \mapsto \mathcal{O}(f) = \int \mathcal{O}(x)\, f(x) \mathrm{d}^{d+1}x \in \mathcal{L}(\mathcal{D}, \mathcal{H}_0) \right\} \tag{4.99}$$

defined on a common, dense, invariant, domain $\mathcal{D}$ given by the subspace of vectors $|\chi\rangle \in \mathcal{H}_0$ such that $\|(1 + P^0)^\ell \chi\| < \infty$ for all $\ell > 0$. These operator valued distributions are unbounded but each field in this collection satisfies a bound of the form $\|(1 + P^0)^{-\ell} \mathcal{O}(x)(1 + P^0)^{-\ell}\| < \infty$ for some sufficiently large number ℓ, i.e. field operators exist point-wise if we damp them appropriately. The smeared fields $\mathcal{O}(f)$ associated with test-functions $f \in C_0^\infty(O)$ localized in a causal diamond O are "affiliated" with the local v. Neumann algebra $\pi_0(\mathfrak{A}(O))''$ in the sense that their spectral projections are elements of this algebra (They cannot of course themselves be in the local v. Neumann algebra because they are unbounded). The fields $\mathcal{O}$ can furthermore be arranged into multiplets transforming naturally under Poincaré transformations in the sense that $U(\Lambda, a)\mathcal{O}(x)U(\Lambda, a)^* = D(\Lambda^{-1})\mathcal{O}(\Lambda x + a)$, where D is some irreducible representation of the covering of the Lorentz group $SO_+(d, 1)$. We will assume for simplicity that all of these representations D are finite dimensional, i.e. that the multiplets have a finite number of components.

If the underlying net is even conformally invariant in the sense of (a3′), (a4′), then it is natural to assume that the fields can be suitably re-organized into (larger) multiplets of the conformal group $\widetilde{G}$. By this we shall mean that among all fields $\mathcal{O} \in \Phi$ there is a countable subset of linearly independent "primary fields". These by definition should transform as (in the sense of operator-valued distributions) [30]

$$U(g)\mathcal{O}(x)U(g)^* = N(g, x)^{d_\mathcal{O}} D[\Lambda(g, x)^{-1}]\mathcal{O}(g \cdot x) \tag{4.100}$$

for all pairs (x, g) of points x and conformal transformations $g \in \widetilde{G}$ such that $g \cdot x$ can be deformed to x for a path of conformal transformations $t \mapsto g(t)$ without passing through the point at infinity. Here, $N(g, x)$ is the conformal factor of the transformation, i.e. $N(g, x)^2 \eta_{\mu\nu} = g^* \eta_{\mu\nu}$. $d_{\mathcal{O}} \geq 0$ is called the "dimension" of the primary field. $D(\Lambda(g, x))$ implements the tensorial transformation behavior of the field, where $\Lambda(g, x) = N(g, x)^{-1} \partial(g \cdot x)/\partial x \in \mathrm{SO}_+(d, 1)$ is a Lorentz transformation associated with g, x, and D is an irreducible, finite dimensional, representation of $\mathrm{SO}_+(d, 1)$, see [24, 30] for more explicit expressions. Besides primary fields, there are "descendants", which are by definition fields of the form $\mathcal{O}_{\mu_1 \dots \mu_k} = [P_{\mu_1}, [\dots, [P_{\mu_k}, \mathcal{O}]]]$, where $\mathcal{O}$ is a primary field. The dimension of such a descendant is then defined to be $d_{\mathcal{O}} + k$. We assume that the set of all fields Φ is spanned by the countably many primary fields and their (countably many) descendant fields.[10] We also assume that

$$\mathrm{span}\left\{\mathcal{O}(f)|0\rangle \mid f \in C_0^\infty, \mathcal{O} \in \Phi\right\} \text{ is dense in } \mathcal{H}_0, \qquad (4.101)$$

i.e. we may approximate in norm, with arbitrary precision, any vector in $\mathcal{H}_0$ by applying a suitable combination of smeared field $\mathcal{O}(f)$ to the vacuum. Under these assumptions we now show:

Theorem 14 *Let A be a double cone whose base is a ball of radius r, and let B be the causal complement of a double cone whose base is a concentric ball of radius $R > r$. Let ω_0 be the vacuum state. Under the assumptions on our conformal field theory just described, we have*

$$E_R(\omega_0) \leq \ln \sum_{\mathcal{O} \in \Phi} \left(\frac{r}{R}\right)^{d_{\mathcal{O}}}. \qquad (4.102)$$

Remark 6 A corollary of the theorem is that if $\mathcal{O}$ is the primary field with the smallest non-zero dimension $d_{\mathcal{O}}$, then for large $R \gg r$, we have

$$E_R(\omega_0) \lesssim N_{\mathcal{O}} \left(\frac{r}{R}\right)^{d_{\mathcal{O}}} \qquad (4.103)$$

where $N_{\mathcal{O}}$ is the number of independent components of $\mathcal{O}$.

Example For a free hermitian massless scalar field ϕ in $3 + 1$ dimensions, a basis of fields $\mathcal{O}$ for the set Φ can be chosen to be the Wick monomials $\mathcal{O}(x) =: D^{(n_1)}\phi(x) \cdots D^{(n_k)}\phi(x) :$, where the double dots denote normal ordering, i.e. all creation operators are put to the left of all annihilation operators upon inserting relation (2.36). The derivative operators $D^{(n)}$ are defined as

---

[10]It would be interesting to see whether such an assumption can be derived from the basic axioms (a1), (a2), (a3′), (a4′), (a5′). Partial progress in this direction has been made by [31].

$$D^{(n)}_{\mu_1...\mu_n} = P^{\nu_1...\nu_n}_{\mu_1...\mu_n} \partial_{\nu_1} \cdots \partial_{\nu_n}, \tag{4.104}$$

with $P^{\nu_1...\nu_n}_{\mu_1...\mu_n}$ denoting the projection onto tensors which are trace free with respect to any pair of indices (upon contraction with $\eta^{\mu_i \mu_j}$). The trace free condition arises from the fact that $\partial^\mu \partial_\mu \phi = 0$. The dimension is given by $d_{\mathcal{O}} = k + n_1 + \cdots + n_k$. The dimension of the space of trace free tensors of rank n is given by $(n + 1)^2$ in $3 + 1$ spacetime dimensions. From this, the conformal partition function is found to be

$$\ln \sum_{\mathcal{O} \in \Phi} \left(\frac{r}{R} \right)^{d_{\mathcal{O}}} = \ln \prod_{n=1}^{\infty} \left(\frac{1}{1 - (r/R)^n} \right)^{n^2} \lesssim \frac{\pi^4}{45} \tau^{-3}, \tag{4.105}$$

as $\tau \to 0^+$, where $r/R = e^{-\tau}$, so according to our theorem $E_R(\omega_0) \lesssim \frac{\pi^4}{45} \tau^{-3}$ as $r \to R$. On the other hand, the field with the smallest dimension which is not the identity is ϕ itself, and $d_\phi = 1$. From this one finds $E_R(\omega_0) \lesssim r/R$ for $R \gg r$.

Proof By conformal invariance, we may again assume without loss of generality that $R = 1 > r > 0$. The idea of the proof is to define the vectors

$$|\mathcal{O}\rangle = \Delta^{\frac{1}{4}} \mathcal{O}(0)|0\rangle, \tag{4.106}$$

where $\mathcal{O}$ runs through some basis of Φ, and where here and in the rest of the proof, $\Delta = \Delta_{B'}$ is the modular operator for the region $O_{B'} = O'_B$. Evaluating the operator identity of Lemma 13 for $t = -i/4$ formally gives for $s > 0$

$$\exp\left[-\frac{s}{2}(P^0 + K^0) \right]|\mathcal{O}\rangle = \exp\left[-\frac{s}{2}(P^0 + K^0) \right] \Delta^{\frac{1}{4}} \mathcal{O}(0)|0\rangle = \Delta^{\frac{1}{4}} e^{isD} \mathcal{O}(0)|0\rangle. \tag{4.107}$$

On the other hand, from the relation $e^{isD} P_\mu e^{-isD} = e^{-s} P_\mu$ of the conformal algebra, the fact that the conformal factor for a dilation by λ is λ, and the invariance of the vacuum, we have $e^{isD} \mathcal{O}(0)|0\rangle = e^{isD} \mathcal{O}(0)e^{-isD}|0\rangle = e^{-sd_\phi} \mathcal{O}(0)|0\rangle$ for any primary or descendant field $\mathcal{O}$. It follows that

$$\exp\left[-\frac{s}{2}(P^0 + K^0) \right]|\mathcal{O}\rangle = e^{-sd_{\mathcal{O}}}|\mathcal{O}\rangle. \tag{4.108}$$

If we can show that $\{|\mathcal{O}\rangle \mid \mathcal{O} \in \Phi\}$ forms a basis of $\mathcal{H}_0$, then the vectors with fixed $d_{\mathcal{O}}$ span an eigenspace of $\exp\left[-\frac{s}{2}(P^0 + K^0) \right]$. Putting $s = -\ln(r)$ we therefore find

$$\text{Tr } r^{\frac{1}{2}(P^0 + K^0)} = \sum_{\mathcal{O} \in \Phi} r^{d_{\mathcal{O}}}. \tag{4.109}$$

In view of Theorem 13, this would complete the proof. We now make the above somewhat formal arguments rigorous. In order to do this, we first note that $\mathcal{O}(x)|0\rangle$ is a $\mathcal{H}_0$-valued distribution that is the boundary value of a strongly holomorphic $\mathcal{H}_0$-valued (see Appendix A) function in the domain $\mathbb{R}^{d,1} + iV^+$, where V^+ is the interior of the future lightcone. Indeed, this holomorphic extension may be defined as (here $Pz = P_\mu z^\mu$)

$$\mathcal{O}(z)|0\rangle = e^{-iPz}\mathcal{O}(0)|0\rangle := e^{-iPz}(1+P^0)^\ell \cdot [(1+P^0)^{-\ell}\mathcal{O}(x)(1+P^0)^{-\ell}]|0\rangle, \tag{4.110}$$

noting that

$$\|\mathcal{O}(z)|0\rangle\| \leq \|e^{-iPz}(1+P^0)^\ell\| \, \|(1+P^0)^{-\ell}\mathcal{O}(x)(1+P^0)^{-\ell}\| \leq C_\ell(\Im(z^0) - |\Im(\mathbf{z})|)^{-\ell}. \tag{4.111}$$

Hence, by a simple generalization of Theorem 3.1.15 [32] to Hilbert-space valued distributions, $\mathcal{O}(x)|0\rangle$ is indeed the distributional boundary value in the strong sense[11] of the holomorphic function $\mathbb{R}^{d,1} + iV^+ \ni z \mapsto \mathcal{O}(z)|0\rangle \in \mathcal{H}_0$. Next we use again the Hislop-Longo theorem [26] stating that the modular group $\Delta_{B'}^{it}$ is equal to $U(L(t))$, where $L(t)$ was given above by (4.93). Let $x(t) = L(t) \cdot x$ (for $x \in O_{B'}$), and let $x_\pm = x^0 \pm |\mathbf{x}|$. Then it follows that

$$x_\pm(t) = \frac{(1+x_\pm) - e^{2\pi t}(1 - x_\pm)}{(1+x_\pm) + e^{2\pi t}(1 - x_\pm)} \tag{4.113}$$

while $\mathbf{x}(t)/|\mathbf{x}(t)| = \mathbf{x}/|\mathbf{x}|$ for all $t \in \mathbb{R}$. It is easy to check from this expression that, for fixed $t_0 > 0$ and x in a sufficiently small neighborhood O of the origin, the complex points $x(t - is)$, $0 < s < \frac{1}{4}$, $|t| < t_0$ remain within $\mathbb{R}^{d,1} + iV_+$. Thus, $\mathcal{O}(x(t - is))|0\rangle$ is a well-defined vector in $\mathcal{H}_0$ for all $x \in O$. The conformal factor $N(x,t) \equiv N(L(t), x)$, and the associated Lorentz transformation, $\Lambda(x,t) \equiv \Lambda(L(t), x)$ of the conformal transformations $L(t)$ (4.93) appearing in the transformation law (4.100) are found to be[12]

$$N(x,t) = \left(\cosh(\pi t) - x_+ \sinh(\pi t)\right)^{-1}\left(\cosh(\pi t) - x_- \sinh(\pi t)\right)^{-1},$$
$$\Lambda(x,t) = \exp\left(\frac{2\mathbf{x} \cdot \mathbf{C}}{x_+ - x_-} \ln \frac{\cosh(\pi t) - x_+ \sinh(\pi t)}{\cosh(\pi t) - x_- \sinh(\pi t)}\right), \tag{4.114}$$

where $\mathbf{C} = (M_{01}, \ldots, M_{0d})$ are the generators of boosts in $\mathfrak{so}(d+1, 2)$. It can be seen from these expressions that the analytic continuation $N(x, t - is)$ avoids the negative real axis for $|t| < t_0$, $0 \leq s < \frac{1}{4}$ as long as $x \in O$ and as long as O is a sufficiently small neighborhood of the origin (depending on t_0). Similarly, $\Lambda(x, t - is)$ remains single-valued in this range. These facts imply that, for a primary field $\mathcal{O}$,

---

[11] This means that

$$\mathcal{O}(f)|0\rangle = \lim_{y \to 0, y \in V^+} \int d^{d+1}x \, \mathcal{O}(x + iy)|0\rangle f(x). \tag{4.112}$$

where the limit is understood in the norm topology on $\mathcal{H}_0$.

[12] The second relation can be found by integrating the explicit infinitesimal versions of the transformation law given e.g. in [24, 30].

$$\Delta^{s+it}\mathcal{O}(x)|0\rangle = N(x, t - is)^{d_\mathcal{O}} D[\Lambda(x, t - is)^{-1}]\mathcal{O}(x(t - is))|0\rangle \qquad (4.115)$$

pointwise for all $x \in O$, $|t| < t_0$, $0 < s < \frac{1}{4}$, as both sides have the same distributional boundary value (in x) when $s \to 0^+$ by the transformation law (4.100) for primary fields, and hence must coincide by the edge-of-the-wedge theorem, see Appendix A.1. We may now set in this equation $x = 0, t = 0$ and let $s \to \frac{1}{4}^-$. Using $x_\pm(-\frac{1}{4}i) = i$, we find $x(-\frac{1}{4}i) = ie_0 = (i, 0, 0 \dots, 0)$, $\Lambda(0, -\frac{1}{4}i) = 1$ and $N(0, -\frac{1}{4}i) = 1$, and we arrive at the formula

$$|\mathcal{O}\rangle = \Delta^{\frac{1}{4}}\mathcal{O}(0)|0\rangle = \mathcal{O}(ie_0)|0\rangle. \qquad (4.116)$$

Since e_0 is clearly inside the forward lightcone, the right side is a well-defined, non-zero vector in $\mathcal{H}_0$ (finite norm). Thus, we have shown that $|\mathcal{O}\rangle$ is a well-defined vector when $\mathcal{O}$ is a primary field. By applying a suitable number of commutators with $[P_\mu, \; . \;]$ to (4.115), we can easily reach a similar conclusion for descendant fields $\mathcal{O}_{\mu_1 \dots \mu_k} = [P_{\mu_1}, [\dots, [P_{\mu_k}, \mathcal{O}]]]$, namely

$$|\mathcal{O}_{\mu_1 \dots \mu_k}\rangle = i^k \, \partial_{\mu_1} \dots \partial_{\mu_k} [N(x; -\tfrac{1}{4}i)^{d_\mathcal{O}} D[\Lambda(x, -\tfrac{1}{4}i)^{-1}] \, \mathcal{O}(x(-\tfrac{1}{4}i))]|0\rangle \Big|_{x=0}.$$
$$(4.117)$$

That the set $\{|\mathcal{O}\rangle \mid \mathcal{O} \in \Phi\}$ forms a basis of $\mathcal{H}_0$ can now be seen as follows. Assume that $|\chi\rangle$ is orthogonal to all $|\mathcal{O}\rangle$, $\phi \in \Phi$. It follows from (4.117) that for any k, we have $\langle\chi|(\partial_{\mu_1} \dots \partial_{\mu_k}\mathcal{O})(ie_0)|0\rangle = 0$. Since $\mathcal{O}(z)|0\rangle$ is holomorphic in an open neighborhood in $\mathbb{C}^{d+1}$ of $z = ie_0$, it follows that $\langle\chi|\mathcal{O}(z)|0\rangle = 0$ for all z in such a neighborhood. By the edge-of-the-wedge theorem, it follows that $\langle\chi|\mathcal{O}(x)|0\rangle = 0$ in the distributional sense (i.e. after smearing with $f(x)$). Since this holds for all fields $\mathcal{O} \in \Phi$, we conclude that $|\chi\rangle$ is in the orthogonal complement of the set (4.101). Since that set is by assumption dense in $\mathcal{H}_0$, we conclude that $|\chi\rangle = 0$, i.e. we learn that $\{|\mathcal{O}\rangle \mid \mathcal{O} \in \Phi\}$ spans a dense subset of $\mathcal{H}_0$.

We next show that the elements in the set $\{|\mathcal{O}\rangle \mid \mathcal{O} \in \Phi\}$ are linearly independent. Suppose that there exists a vanishing finite linear combination $\sum_i c_i|\mathcal{O}_i\rangle = 0$ for a set of linearly independent fields $\mathcal{O}_i \in \Phi$. Using (4.117), we can also write this as $\sum_i c_i'\mathcal{O}_i'(ie_0)|0\rangle = 0$, where c_i' is a new set of complex numbers and $\mathcal{O}_i'$ a new set of linearly independent fields in Φ. Let $\psi = \sum_i c_i'\mathcal{O}_i'$. We conclude that $\psi(x + i\varepsilon e_0)|0\rangle = e^{-iP \cdot x} e^{(1-\varepsilon)P^0}\psi(ie_0)|0\rangle = 0$ for any sufficiently small $\varepsilon > 0$ and all $x \in \mathbb{R}^{d,1}$. Thus, by the edge-of-the-wedge theorem (see Appendix A.1), $f \mapsto \psi(f)|0\rangle = 0$ in the distributional sense. Since $\psi(f)$ is affiliated with $\mathfrak{A}(O)$, we have $[\psi(f), a] = 0$ in the strong sense for $a \in \mathfrak{A}(O')$, so $\psi(f)a|0\rangle = 0$. By the Reeh-Schlieder theorem, the set $a|0\rangle, a \in \mathfrak{A}(O')$ is dense in $\mathcal{H}_0$, therefore we see that $\psi(f) = 0$ for all test functions f, or in other words, $\sum_i c_i'\mathcal{O}_i' = 0$ as an identity between quantum fields (i.e. when smeared with any test function f). Thus, we see that $c_i' = 0$, and this is also easily seen to imply that all $c_i = 0$. This completes the proof that $\{|\mathcal{O}\rangle \mid \mathcal{O} \in \Phi\}$ forms a basis of $\mathcal{H}_0$.

The rest of the argument leading to Eq. (4.109) can now also be made rigorous using Eq. (4.115) repeating the above formal steps with this equation for $0 < s < \frac{1}{4}$, and taking $s \to \frac{1}{4}^-$ in the end. $\qquad\qquad\qquad\qquad\qquad\qquad\qquad\qquad\qquad\qquad\qquad$ □

4.5 Upper Bounds for CFTs in 3 + 1 Dimensions

In the previous section, we have treated the case when O_A is a diamond whose base is a ball, A, and where O_B is the complement of a concentric diamond. It is of interest to obtain also a bound when the diamonds are in arbitrary position (i.e. not concentric), but still of course $O_A \subset O'_B$ is in the causal complement, see Fig. 1.3. Upper bounds can be obtained in this case by essentially the same method as in the previous subsection, but the formula for the upper bound becomes somewhat more complicated. To keep the complications at a minimum, we will only consider the case when $d = 3$, i.e. $3 + 1$ dimensional CFTs.

The key point is basically to understand the finite dimensional irreducible representations of $\widetilde{SO_+(3, 1)} \cong SL_2(\mathbb{C})$, D, in the transformation formula (4.100) for the quantum fields. These are best described in spinorial form. The inequivalent D's are labelled by two natural numbers s, s' and act on the vector space

$$V_{s,s'} = E_s(\mathbb{C}^{2\,\otimes s}) \otimes E_{s'}(\bar{\mathbb{C}}^{2\,\otimes s'}) \tag{4.118}$$

where E_s projects onto the subspace of symmetric rank s tensors. The action of $D_{s,s'}(g), g \in SL_2(\mathbb{C})$ on a tensor $T \in V_{s,s'}$ is given by

$$(D_{s,s'}(g)T)_{A_1\ldots A_s B'_1\ldots B'_{s'}} = g_{A_1}{}^{C_1} \cdots g_{A_s}{}^{C_s} \bar{g}_{B'_1}{}^{D'_1} \cdots \bar{g}_{B'_{s'}}{}^{D'_{s'}} T_{C_1\ldots C_s D'_1\ldots D'_{s'}}. \tag{4.119}$$

Tensors over $\mathbb{R}^{3,1}$ correspond to elements of $V_{s,s'}$ by the rules explained in detail e.g. in [33]. For instance, an anti-symmetric tensor $T_{\mu\nu} = -T_{\nu\mu}$ decomposes into one complex component T_{AB} in $V_{2,0}$ and another one $\bar{T}_{A'B'}$ in $V_{0,2}$. The "spin" of the finite dimensional representation is $S = \frac{1}{2}s + \frac{1}{2}s'$, and we can also define the left and right chiral spins by $S^L = \frac{1}{2}s$, $S^R = \frac{1}{2}s'$. The transformation behavior (4.100) of a quantum field $\mathcal{O}$ under Lorentz transformations is described by $S_\mathcal{O}^L, S_\mathcal{O}^R$, and the transformation behavior under dilations by its dimension, $d_\mathcal{O}$. We can now state our result.

Theorem 15 *Let O_A be a double cone which is the intersection of the past of a point x_{A+} and the future of a point x_{A-}. Similarly, let O_B be the complement of a double cone which is the intersection of the past of a point x_{B+} and the future of a point x_{B-}. It is required that O_A is properly contained in the other double cone, i.e. the causal complement of O_B. Define the conformally invariant cross-ratios by*

$$u = \frac{(x_{B+} - x_{B-})^2 (x_{A+} - x_{A-})^2}{(x_{A-} - x_{B-})^2 (x_{A+} - x_{B+})^2} > 0$$

$$v = \frac{(x_{B+} - x_{B-})^2 (x_{A+} - x_{A-})^2}{(x_{A-} - x_{B+})^2 (x_{A+} - x_{B-})^2} > 0,$$

(4.120)

and let τ, θ *be defined by*

$$\theta = \cosh^{-1}\left(\frac{1}{\sqrt{v}} - \frac{1}{\sqrt{u}}\right), \quad \tau = \cosh^{-1}\left(\frac{1}{\sqrt{v}} + \frac{1}{\sqrt{u}}\right). \tag{4.121}$$

Let ω_0 *be the vacuum state. Under the assumptions on our conformal field theory described in the previous subsection, we have in* $3 + 1$ *dimensions:*

$$E_R(\omega_0) \leq \ln \sum_{\mathcal{O} \in \Phi}' e^{-\tau d_\mathcal{O}} [2S_\mathcal{O}^R + 1]_\theta [2S_\mathcal{O}^L + 1]_\theta, \tag{4.122}$$

with $[n]_\theta = (e^{n\theta/2} - e^{-n\theta/2})/(e^{\theta/2} - e^{-\theta/2})$. *For* $\tau \sim |\theta| \gg 1$ *this gives*

$$E_R(\omega_0) \lesssim N_\mathcal{O} \cdot e^{-\tau(d_\mathcal{O} - S_\mathcal{O})}, \tag{4.123}$$

where $\mathcal{O}$ *is the operator with the smallest "twist"* $d_\mathcal{O} - S_\mathcal{O}$ *and* $N_\mathcal{O}$ *its multiplicity.*

Remark 7 Note that, unlike in Theorem 14, the sum $\sum'$ over $\mathcal{O}$ is over all different independent field *multiplets* under $SL_2(\mathbb{C})$, not their individual operator *components*. Thus, for instance, a hermitian tensor field operator $\mathcal{O}_{\mu\nu}$ satisfying $\mathcal{O}_{\mu\nu} = -\mathcal{O}_{\nu\mu}$ would correspond under the identification $\varepsilon_{AB}\mathcal{O}_{AB} + \bar{\varepsilon}_{A'B'}\mathcal{O}^*_{A'B'}$ to 2 multiplets, namely $\mathcal{O}_{AB}$ and $\mathcal{O}^*_{A'B'}$, one having $S_\mathcal{O}^L = 1$, $S_\mathcal{O}^R = 0$, and the other $S_\mathcal{O}^L = 0$, $S_\mathcal{O}^R = 1$, and not 6 real component fields.

Proof By applying a conformal transformation to the double cones, one can achieve that

$$x_{B\pm} = \pm(1, 0, 0, 0), \quad x_{A\pm} = (\pm e^{-\tau} \cosh \theta, e^{-\tau} \sinh \theta, 0, 0) \tag{4.124}$$

for some τ, θ satisfying $\tau > |\theta|$ (the last statement uses the assumptions on the relative position of the diamonds). Computing the cross ratios u, v for these points, one finds precisely the relations (4.121). Since the CFT is conformally invariant in the sense of $(a3')$, $(a4')$ it suffices to prove the theorem for this special configuration. As in the previous subsection, let Δ_A, $\Delta_{B'}$ be the modular operators for the diamonds O_A, O'_B. Define the operator T as before in (4.87). As before, it follows that $E_R(\omega_0) \leq$ ln Tr $|T|$. Thus, we need to compute this trace.
    Define

$$P^1 = -\frac{1}{i}\frac{d}{dt}U\left(\exp t(M_{15}+M_{14})\right)\Big|_{t=0},$$

$$K^1 = -\frac{1}{i}\frac{d}{dt}U\left(\exp t(M_{15}-M_{14})\right)\Big|_{t=0}, \qquad (4.125)$$

$$L_{01} = \frac{1}{i}\frac{d}{dt}U\left(\exp t M_{01}\right)\Big|_{t=0},$$

which are the generators of translations/special conformal transformations in the 1-direction, and boosts in the 01-plane. By conformal invariance, we can write (compare (4.95))

$$T = \Delta_{B'}^{\frac{1}{4}}e^{-i\tau D+i\theta L_{01}}\Delta_{B'}^{-\frac{1}{4}}e^{i\tau D-i\theta L_{01}} \equiv Xe^{i\tau D-i\theta L_{01}}. \qquad (4.126)$$

Thus, $\mathrm{Tr}\,|T| = \mathrm{Tr}\,|X|$. In basically the same way as in the previous subsection, one derives the operator identity (compare (4.98))

$$X = \exp\left[-\frac{1}{2}\tau(P^0+K^0)+\frac{1}{2}\theta(P^1-K^1)\right], \qquad (4.127)$$

so $\mathrm{Tr}\,|X| = \mathrm{Tr}\,\exp[-\frac{1}{2}\tau(P^0+K^0)+\frac{1}{2}\theta(P^1-K^1)]$. Define next the vectors $|\mathcal{O}\rangle$ as in (4.106). By the same arguments as there (compare (4.108)), we find

$$X|\mathcal{O}\rangle = e^{-\tau d_\mathcal{O}}D_{s_\mathcal{O},s'_\mathcal{O}}[\exp(-\theta M_{01})]|\mathcal{O}\rangle. \qquad (4.128)$$

We must now determine how the finite dimensional matrix $D_{s_\mathcal{O},s'_\mathcal{O}}(\exp\theta M_{01})$ acts on an operator $\mathcal{O}$ transforming in the representation $V_{s_\mathcal{O},s'_\mathcal{O}}$ of $SL_2(\mathbb{C})$. This is conveniently done by choosing a basis e_0, e_1 in $\mathbb{C}^2$ such that, relative to this basis

$$(D_{s_\mathcal{O},s'_\mathcal{O}}(\exp\theta M_{01})T)_{A_1...A_s B'_1...B'_{s'}} = e^{\frac{1}{2}\theta(n_0-n_1)+\frac{1}{2}\theta(n'_0-n'_1)}T_{A_1...A_s B'_1...B'_{s'}}, \qquad (4.129)$$

where n_0 is the number of times an A_i assumes the value 0, n_1 the number of times an A_i assumes the value 1, where n'_0 is the number of times a B'_i assumes the value 0, and n'_1 the number of times a B'_i assumes the value 1. The above expression follows from the way in which the isomorphism $\widetilde{SO_+(3,1)} \cong SL_2(\mathbb{C})$ is set up. We may now compute the trace $\mathrm{Tr}\,|X|$ in the basis $\{|\mathcal{O}\rangle\}$. A straightforward computation using $n_0+n_1 = s_\mathcal{O}, n'_0+n'_1 = s'_\mathcal{O}$ gives the right side of the formula (4.122), and the proof is complete. $\qquad\qquad\Box$

4.6 Upper Bounds for Chiral CFTs

Here we apply the general upper bounds provided by Theorems 3 and 4 to the case of a chiral CFT described by a net of v. Neumann algebras $\{\mathfrak{A}(I)\}$ over the circle S^1. We consider the entanglement entropy $E_R(\omega_0)$ of two disjoint open intervals $A, B \subset S^1$ in the vacuum state ω_0. By abuse of notation, we denote these intervals by $A = (a_1, a_2)$, $B = (b_2, b_1)$, where[13] $a_1, a_2, b_1, b_2 \in S^1$. We denote the GNS-representation of ω_0 on $\mathcal{H}$ by π_0 and the vacuum vector by $|0\rangle$. The unitary projective positive energy representation of the covering $\widetilde{G}$ of the conformal group $G = \mathrm{SU}(1, 1)$ on $\mathcal{H}$ is denoted by $U_0(g)$, $g \in \widetilde{G}$; invariance of the vacuum means that $U_0(g)|0\rangle = |0\rangle$ for all $g \in \widetilde{G}$. The infinitesimal generator of rotations of S^1 in the representation π_0 is denoted by $L_0 = \mathrm{d}/\mathrm{d}t\, U_0(\mathrm{diag}(e^{it/2}, e^{-it/2}))|_{t=0}$. The following result can be obtained in exactly the same way as those in the previous two subsections.

Theorem 16 *Let ω_0 be the vacuum state. We have*

$$E_R(\omega_0) \le \ln\left(\mathrm{Tr}\exp\left\{-2\sinh^{-1}\sqrt{\xi}\cdot L_0\right\}\right) \qquad (4.130)$$

where ξ is the conformally invariant cross ratio associated with the pair of intervals $A = (a_1, a_2)$, $B = (b_2, b_1) \subset S^1$ given by

$$\xi = \frac{(a_2 - b_2)(b_1 - a_1)}{(a_2 - a_1)(b_2 - b_1)}. \qquad (4.131)$$

Remark 8 (1) The upper bound is obviously only non-trivial if $e^{-\tau L_0}$ has finite trace for $\tau > 0$, which is the case e.g. in all rational conformal field theories. In many such theories, there are concrete formulas for the character $\mathrm{Tr}\, e^{-\tau L_0}$ leading to an explicit bound under the substitution $\tau = 2\sinh^{-1}\sqrt{\xi}$, see e.g. [34].

(2) As before in Theorems 13 and 15, our bound is an upper bound on $E_M(\omega_0) \ge E_R(\omega_0)$ as well, where E_M is the modular entanglement measure.

Proof Define again an operator T as above in (4.87), so that, as argued in the proof of Theorem 13, $E_R(\omega_0) \le \ln\mathrm{Tr}\,|T^*|$, where $|T^*| = \sqrt{TT^*}$. [25] show

$$\sqrt{TT^*} = e^{-\ell(A,B)L_0}. \qquad (4.132)$$

Here $\ell(A, B)$ is an "inner distance" associated with the inclusion $(a_1, a_2) \subset (b_1, b_2)$. Going through the implicit definitions given by these authors, one finds the explicit formula $\ell(A, B) = 2\sinh^{-1}\sqrt{\xi}$. $\qquad\square$

Via the Cayley transform $z \in S^1 \mapsto i(z-1)/(z+1) \in \mathbb{R}$ we get a corresponding bound for the theory on the light ray $\mathbb{R}$, which, in fact, has the same form (with a_i, b_i now in $\mathbb{R}$ rather than S^1), since the cross ratio ξ retains its form under the Cayley transform.

We can also get asymptotic formulas for the entanglement entropy (e.g. in the light-tray picture) using the known behavior of the character $\mathrm{Tr}\, e^{-\tau L_0}$ for small respectively large $\tau > 0$. These bounds are conveniently expressed in terms of the dimensionless ratio $r = \frac{\mathrm{dist}(A,B)}{\sqrt{|A|\cdot|B|}}$, where $A, B \subset \mathbb{R}$ have lengths $|A|, |B|$. Thus, widely separated

---

[13]We adopt the convention that points on S^1 are labelled clockwise.

intervals have $r \gg 1$ and $r \simeq \sqrt{\xi}$, while intervals separated by a short distance compared to their length have $r \ll 1$. Using that the smallest non-zero eigenvalue of L_0 is 1 with some multiplicity n_1 in the vacuum sector, the theorem immediately implies

$$E_R(\omega_0) \lesssim \frac{n_1}{4r^2} \quad \text{for } r \gg 1. \tag{4.133}$$

We can also get a bound in the opposite regime $r \ll 1$ if we have an asymptotic bound on $\operatorname{Tr} e^{-\tau L_0}$ for $\tau \to 0^+$. For a rational conformal chiral net, [35] have shown for instance that

$$\ln \operatorname{Tr} \, e^{-\tau L_0} = \frac{c\pi^2}{6\tau} - \frac{1}{2} \ln \mu_{\mathfrak{A}} + O(\tau), \tag{4.134}$$

where $\mu_{\mathfrak{A}}$ is the so-called μ-index of the net $\{\mathfrak{A}(I)\}$, given by the sum of the square of the statistical dimensions $\mu_{\mathfrak{A}} = \sum \dim(\rho)^2$ over all irreducible sectors (see the next subsection for further explanations concerning this notion). This implies $E_R(\omega_0) \lesssim \frac{c\pi^2}{6r}$, for $r \ll 1$, which falls short of the expected [36] logarithmic behavior of $E_R(\omega_0)$ for $0 < r \ll 1$. We should remember, however, that our bound is also an upper bound on $E_M(\omega_0)$, since the proof really estimates this quantity and since $E_M \geq E_R$ in general. If we use instead the tighter bound $E_I \geq E_R$ in terms of the mutual information, then using the exact result $E_I(\omega_0) = -\frac{c}{3} \ln \xi$ recently obtained by [37] (building on earlier work by [38]) for free fermions where $c = 1/2$, one gets this logarithmic behavior at least in that case.

4.7 Charged States

According to the general philosophy, the algebras of observables $\mathfrak{A}(O)$ contain the observables of the theory accessible to an observer in O. They are not, however, supposed to contain non-observable fields, such as e.g. charge carrying fields, where "charge" is understood here in a rather broad sense. For instance, in a theory containing fermionic fields, only bosonic combinations (such as bi-linears in such fields) would be in $\mathfrak{A}(O)$. Similarly, fields that are charged under some group G of internal symmetries, would not be in $\mathfrak{A}(O)$, only combinations which are invariant under G, i.e. 'singlets'. In the algebraic approach, these objects arise "through the back door" when considering the GNS-representation of "charged states", or simply "charged representations". It is outside the scope of this article to review this beautiful theory, initiated by [39–41], see [6, 42, 43] for the case of quantum field theories in 1+1 dimensions, where qualitatively new phenomena are possible.

Here we only give the basic framework and make some comments. For more comprehensive expositions besides the original papers see [44–46]. For simplicity, we will work in Minkowski spacetime, and we let π_0 be the vacuum representation. Using π_0, we pass to the corresponding net $\pi_0(\mathfrak{A}(O))''$ of v. Neumann algebras on $\mathcal{H}_0$, which by abuse of notation is denoted again by $\mathfrak{A}(O)$. A localized, charged representation π_ρ is one such that there exists a double cone O such that π_0 is unitarily equivalent to π_ρ when restricted to $\mathfrak{A}(O')$, where O' is the causal complement, i.e.

$\pi_0(a) = V\pi_\rho(a)V^*$ for all $a \in \mathfrak{A}(O')$. In order to make possible a general analysis, one assumes for technical reasons the so-called "Haag duality", i.e. $\mathfrak{A}(O)' = \mathfrak{A}(O')$ [a strengthened version of (a2)] for all causal diamonds, which is satisfied in many models. By identifying the Hilbert spaces $\mathcal{H}_0$ and $\mathcal{H}_\rho$ with the isometry V, one can then easily show that there exists an algebra homomorphism ρ of $\mathfrak{A} = \cup_O \mathfrak{A}(O)$ such that

$$\pi_\rho \circ \rho = \pi_0, \quad \rho(a) = a \quad \text{for all } a \in \mathfrak{A}(O'). \tag{4.135}$$

Since we identify the net with its representation under π_0, the vacuum representation is effectively the identity, and we may thus drop the symbol π_0. Because of the last property, ρ is a "localized endomorphism", i.e. it acts non-trivially only on observables localized within O, from which it follows that $\rho(\mathfrak{A}(O)) \subset \mathfrak{A}(O)$. The study of charged representations is thereby reduced to the study of such localized endomorphisms and the associated inclusions of v. Neumann algebras.

One may ask what this notion of charge has to do with the notion of charge carrying field alluded to above. This is clarified by the famous Doplicher-Roberts (DR)-reconstruction theorem [47]. Its basic content is the following. Assume the number of spatial dimensions d is greater than one. Then there exists a "field net" $\{\mathfrak{F}(O)\}$ represented on a larger Hilbert space $\mathcal{H}$ which decomposes as[14]

$$\mathcal{H} = \bigoplus_{[\rho]} \mathcal{H}'_\rho \otimes \mathcal{H}_\rho, \tag{4.136}$$

and a compact group G acting by automorphisms α_g, $g \in$ G on the field net, such that $\mathfrak{A}(O)$ consists precisely of those elements $F \in \mathcal{F}(O)$ that are invariant under all α_g, i.e. $\alpha_g(F) = F$ for all $g \in$ G. The operators $F \in \mathfrak{F}(O)$ which do not have this property are the "charge carrying fields" localized in O. Each α_g is implemented by a unitary representation of the form $U(g) = \oplus_{[\rho]} U_\rho(g) \otimes 1_{\mathcal{H}_\rho}$, where each U_ρ is an irreducible, unitary representation of G on $\mathcal{H}'_\rho$. The charged vectors for irreducible ρ correspond to the vectors in the subspace $\mathcal{H}'_\rho \otimes \mathcal{H}_\rho$, which is often called the "superselection sector" (of ρ).

DR [47] have shown that for each localized endomorphism ρ in O the field algebra $\mathcal{F}(O)$ contains a copy of the Cuntz algebra $\mathcal{O}_{\dim(\rho)}$ (see Sect. 2.2.3), where $\dim(\rho) = \dim(\mathcal{H}'_\rho)$ is called the "statistical dimension" of ρ. If F_i, $i = 1, \ldots, \dim(\rho)$ is a collection of operators ('multiplet') in $\mathcal{F}(O)$ transforming under U_ρ, i.e. $\alpha_g(F_i) = \sum_j U_\rho(g)_{ij} F_j$, then it can be shown that F_i can be written as $F_i = a\psi_i$, where $a \in \mathfrak{A}(O)$ and ψ_i is the generator of a Cuntz algebra sitting inside $\mathcal{F}(O)$. For $a \in \mathfrak{A}$, the action of the endomorphism ρ is

$$\rho(a) = \sum_{i=1}^{\dim(\rho)} \psi_i a \psi_i^*. \tag{4.137}$$

---

[14]Here $[\rho]$ is the equivalence class of all ρ under the natural notion of unitary equivalence. It is meant that there is one summand for each class of irreducible ρ's.

Thus, if ω is a state represented by a vector $|\Psi\rangle$ in $\mathcal{H}_0$, it has no net charge. The state $\omega \circ \rho \equiv \rho^* \omega$ corresponds to adding one unit of charge and can be represented by the vector $\sqrt{\dim(\rho)}\psi_i^*|\Psi\rangle$. In $d = 1$ spatial dimensions, the DR reconstruction theorem does not necessarily hold. However, the notion of localized endomorphism still makes sense. Indeed, one of the most attractive features of the DHR theory of superselection sectors is that the theory can be formulated intrinsically in terms of these. In particular, even in $d = 1$, one can still give an intrinsic definition of the statistical dimension $\dim(\rho) > 0$, which still has many properties of a "dimension", even though it no longer needs to be an integer.

After this brief review, we now make a connection between statistical dimensions and entanglement entropies. As before, we set $\mathfrak{A}_A = \pi_0(\mathfrak{A}_A(O_A))''$ and similarly for B. By abuse of language we say that ρ is localized in A or B if ρ is an endomorphism localized in O_A or O_B. We then have $\rho(\mathfrak{A}_A) \subset \mathfrak{A}_A$, and similarly for B.

Proposition 8 *Let ω be any faithful normal state in the vacuum representation, and $\rho = \prod_i \rho_i^{n_i}$ a product of finitely many irreducible sectors ρ_i with statistical dimensions $\dim(\rho_i)$ localized in A or in B, so that $\rho^* \omega = \omega \circ \rho$ can be thought of as containing for each i precisely n_i additional units of charge of type $[\rho_i]$ relative to ω. Then*

$$0 \leq E_R(\omega) - E_R(\rho^* \omega) \leq \ln \prod_i \dim(\rho_i)^{2n_i}. \tag{4.138}$$

Remark 9 (1) A state is faithful if $\omega(a) = 0$ for $a \in \mathfrak{A}^+$ implies $a = 0$. By the same argument as in the Reeh-Schlieder theorem, this will hold for instance if ω is implemented by a vector in $\mathcal{H}_0$ with finite energy.
(2) Our formula reminds one of results by [48, 49], where the difference between the v. Neumann entropy for A of the vacuum state and a state obtained by applying a charge-carrying field to the vacuum is computed.

Proof To save writing, we put $\mathfrak{M} = \mathfrak{A}_A \vee \mathfrak{A}_B \cong \mathfrak{A}_A \otimes \mathfrak{A}_B$, and $\mathfrak{N} = \rho(\mathfrak{M})$. We denote by σ' a separable state on $\mathfrak{N}$, and by ω' the restriction of ω to $\mathfrak{N}$. By (e4) it immediately follows that $E_R(\omega') \leq E_R(\omega)$. In order to get a lower bound on $E_R(\omega')$ we recall the so-called "left-inverse" of ρ, which is a standard ingredient in DHR-theory [39, 40]. The left inverse is a linear map $\Psi_\rho : \mathfrak{M} \to \mathfrak{M}$ such that $\Psi_\rho \rho = id$ and such that $\Psi_\rho(\rho(a)b\rho(c)) = a\Psi_\rho(b)c$ (it is not an endomorphism in general). A canonical (called "standard") left inverse always exists if ρ is irreducible. [In case we are in > 1 spatial dimensions the DHR reconstruction theorem applies, as described above. In terms of the Cuntz-generators ψ_i, $i = 1, \ldots, \dim(\rho)$, the standard left inverse is then given by the explicit formula $\Psi_\rho(a) = \dim(\rho)^{-1} \sum_i \psi_i^* a \psi_i$, where $\dim(\rho) \in \mathbb{N}$ is the statistical dimension. The left inverse property then follows manifestly from (4.137) and the relations of the Cuntz algebra (2.24).] In that case $\mathcal{E} = \rho \Psi_\rho$ is shown to be a faithful conditional expectation from $\mathfrak{M} \to \mathfrak{N}$. The smallest constant $c > 0$ such that $\mathcal{E}(a^*a) \geq c^{-1}a^*a$ is the Jones-index $[\mathfrak{M} : \mathfrak{N}]$ associated with the inclusion $\mathfrak{N} \subset \mathfrak{M}$ (Pimsner-Popa inequality [50]). By the index-statistics theorem [42, 43],

$$[\mathfrak{M} : \mathfrak{N}]^{\frac{1}{2}} = \dim(\rho), \tag{4.139}$$

so $\mathcal{E}^*\omega \geq \dim(\rho)^{-2}\omega$, implying in particular that $\mathcal{E}^*\omega$ is faithful.

For an arbitrary $\varepsilon > 0$ let σ' be a separable state on $\mathfrak{N}$ such that $E_R(\omega') \geq H(\omega', \sigma') - \varepsilon$. Then $\sigma := \mathcal{E}^*\sigma'$ is a separable state on $\mathfrak{M}$ because, due to the localization properties of Ψ_ρ, $\mathcal{E}$ preserves tensor products in the sense that $\mathcal{E}(a \otimes b) = \mathcal{E}(a) \otimes \mathcal{E}(b)$ for $a \in \mathfrak{A}_A, b \in \mathfrak{A}_B$. Thus, due to the infimum in the definition of $E_R(\omega)$, we get $E_R(\omega) \leq H(\omega, \sigma)$. On the other hand, using the chain rule for the Connes-cocycle (and the definitions of σ' and ω')

$$\begin{aligned}
[D\omega : D\sigma]_t &= [D\omega : D(\mathcal{E}^*\omega)]_t[D(\mathcal{E}^*\omega) : D\sigma]_t \\
&= [D\omega : D(\mathcal{E}^*\omega)]_t[D(\mathcal{E}^*\omega') : D(\mathcal{E}^*\sigma')]_t \quad (4.140) \\
&= [D\omega : D(\mathcal{E}^*\omega)]_t[D\omega' : D\sigma']_t,
\end{aligned}$$

using in the last line the fact that $[D(\mathcal{E}^*\omega') : D(\mathcal{E}^*\sigma')]_t = [D\omega' : D\sigma']_t$, which follows since there exists a faithful, $\mathcal{E}$ invariant state (namely $\mathcal{E}^*\omega$) on $\mathfrak{M}$, see e.g. Sect. 4 of [51] for a discussion. We get

$$\begin{aligned}
H(\omega, \sigma) &= \lim_{t \to 0} \frac{\omega([D\omega : D\sigma]_t - 1)}{it} \\
&= \lim_{t \to 0} \frac{\omega([D\omega' : D\sigma']_t - 1)}{it} + \lim_{t \to 0} \frac{\omega([D\omega : D(\mathcal{E}^*\omega)]_t - 1)}{it} \\
&\quad + \lim_{t \to 0} \frac{\langle([D\omega : D(\mathcal{E}^*\omega)]_t - 1)^*\Omega|([D\omega' : D\sigma']_t - 1)\Omega\rangle}{it} \quad (4.141) \\
&= \lim_{t \to 0} \frac{\omega'([D\omega' : D\sigma']_t - 1)}{it} + \lim_{t \to 0} \frac{\omega([D\omega : D(\mathcal{E}^*\omega)]_t - 1)}{it} \\
&= H(\omega', \sigma') + H(\omega, \mathcal{E}^*\omega) \leq H(\omega', \sigma') + \ln \dim(\rho)^2.
\end{aligned}$$

In the first step we used the alternative definition of H in terms of the Connes-cocycle (91). In the second step we used the chain rule for the Connes-cocycle and (4.140). In the third step we used that $[D\omega' : D\sigma']_t \in \mathfrak{N}$ so that $\omega([D\omega' : D\sigma']_t) = \omega'([D\omega' : D\sigma']_t)$, as well as

$$\begin{aligned}
&\left| \frac{1}{it} \langle([D\omega : D(\mathcal{E}^*\omega)]_t - 1)^*\Omega|([D\omega' : D\sigma']_t - 1)\Omega\rangle \right| \\
&\leq \frac{1}{|t|} \|([D\omega : D(\mathcal{E}^*\omega)]_t - 1)^*\Omega\| \, \|([D\omega' : D\sigma']_t - 1)\Omega\| \quad (4.142) \\
&= 2\left\{ \frac{\mathfrak{R}\omega([D\omega : D(\mathcal{E}^*\omega)]_t - 1)}{t} \frac{\mathfrak{R}\omega'([D\omega' : D\sigma']_t - 1)}{t} \right\}^{\frac{1}{2}} \\
&\to 0 \quad \text{as } t \to 0,
\end{aligned}$$

since $\omega([D\omega : D(\mathcal{E}^*\omega)]_t - 1)/t \to iH(\omega, \mathcal{E}^*\omega)$ and $\omega'([D\omega' : D\sigma']_t - 1)/t \to iH(\omega', \sigma')$ and since $\infty > H(\omega', \sigma')$. In the fourth step, we used again the definition of H in terms of the Connes-cocycle. In the last step we used $\mathcal{E}^*\omega \geq \dim(\rho)^{-2} \, \omega$, the monotonicity of the relative entropy in the second entry, (h5), and $H(\omega, \omega) = 0$. We therefore get

$$E_R(\omega) \leq E_R(\omega') + \varepsilon + \ln \dim(\rho)^2. \tag{4.143}$$

From (h2), one also knows that $E_R(\omega') = E_R(\rho^*\omega)$ (since $\rho : \mathfrak{M} \to \mathfrak{N}$ is faithful). Since ε was arbitrary, the proof is complete for irreducible ρ. In case $\rho = \prod \rho_i^{n_i}$, we proceed by iterating the above argument treating the irreducible endomorphisms ρ_i in the product one by one from the right.   □

Example (Real N-component free KG-field in 3+1 dimensions). The quantum field theory is a simple variant of a 1-component KG theory, the algebraic formulation of which has been described in Sect. 2.4.1. The symplectic space $K_\mathbb{R}$ for the theory with one component is replaced now by N copies $K_\mathbb{R}^N = K_\mathbb{R} \oplus \cdots \oplus K_\mathbb{R}$ corresponding to the N components of the field, i.e. the smearing functions f now have N components $f = (f_I)_{I=1,\dots,N}$. The vacuum state ω_0 and its GNS-triple $(\mathcal{H}, \pi, |0\rangle)$ are only modified in a trivial way. The field algebra (in the sense described above) is $\mathfrak{F}(O) = \pi(\{W(f) \mid \text{supp}(f) \subset O\})''$. An element $g \in O(N)$ acts on a test function by $(g.f)_I = \sum_J g_{IJ} f_J$, and this gives a symplectic map on $K_\mathbb{R}^N$. By the general theory of the Weyl algebra, it corresponds to an automorphism on the field net characterized by $\alpha_g(W(f)) = W(g.f)$. The Hilbert space $\mathcal{H}$ on which the field net acts is the Fock-space of the standard vacuum and it carries a unitary representation $g \mapsto U(g)$ of $O(N)$ implementing α_g in the sense that $U(g)\pi(W(f))U(g)^* = \pi(\alpha_g(W(f)))$. The defining representation π of the field net $\{\mathcal{F}(O)\}$ decomposes as in (4.136), where the labels $[\rho]$ correspond to the irreducible representations of $O(N)$, which in turn are well-known to be characterized by Young tableaux. $\mathcal{H}_0$ is the subspace of $O(N)$ invariant vectors and corresponds to the trivial representation of the net $\{\mathfrak{A}(O)\}$. It is precisely the closure of $\{a|0\rangle \mid a \in \mathfrak{A}(O)\}$ (for any causal diamond O).

Consider now a tensor $T^{I_1 \dots I_k}$ whose symmetry properties under index permutations are characterized by a Young-tableau $\lambda = (\lambda_1, \dots, \lambda_s)$ with k boxes. Next, take functions $f_I \in C^\infty(\mathbb{R}^4)$ with support in a causal diamond O_A with base $A \subset \mathbb{R}^3$ in a time-slice. Define

$$F(T) = \sum_{I_1,\dots,I_k=1}^{N} T^{I_1 \dots I_k} \phi_{I_1}(f_1) \dots \phi_{I_k}(f_k), \tag{4.144}$$

where $\phi_I(f) = \int \phi_I(x) f(x) d^4x$ are the smeared KG quantum fields (so that, $\pi(W(f)) = \exp i \sum_I \phi_I(f_I)$). We assume that our test functions have been chosen so that $F(T) \neq 0$. The transformation law gives $U(g)F(T)U(g)^* = F(g.T)$, where $g.T$ is the action of g on the tensor T. Let $\dim(\lambda)$ be the dimension of this representation and let $\{T_i\}_{i=1,\dots,\dim(\lambda)}$ be an orthonormal basis of tensors with Young-tableau symmetry λ.

By DR theory, there exist corresponding elements $\psi_i \in \mathfrak{F}(O_A), i = 1, \dots, \dim(\lambda)$ satisfying the relations of a Cuntz-algebra and a affiliated with $\mathfrak{A}(O_A)$ such that $F(T_i) = a\psi_i$, and this a can be chosen to satisfy $a^* = a$. ρ defined by (4.137) with $\dim(\rho) = \dim(\lambda)$ is an endomorphism localized in O_A. As one may verify, the corresponding charged state for the net $\{\mathfrak{A}(O)\}$ can be written

$$\rho^* \omega_0(b) = \omega_0(\rho(b)) = \langle \Phi | b | \Phi \rangle, \tag{4.145}$$

where the vector representer of the charged state, $|\Phi\rangle$, is

$$|\Phi\rangle = F(T)^* [\mathcal{F}(F(T)F(T)^*)]^{-\frac{1}{2}} |0\rangle, \tag{4.146}$$

where $\mathcal{F}$ is the mean over the group $K = O(N)$ already defined above in (82). $\dim(\lambda)$ is by the general theory equal to the statistical dimension of the charged state $\rho^* \omega_0$. It is given by a standard formula in terms of the shape of the Young tableau, so we obtain in this example,

$$0 \le E_R(\omega_0) - E_R(\rho^* \omega_0) \le \ln \dim(\lambda)^2 = 2 \ln \prod_{i,j \in \lambda} \frac{(N + j - i)}{h(i, j)}, \tag{4.147}$$

where the "hook length" parameter $h(i, j)$ of a box with coordinates (i, j) (i-th row and j-th column) of the Young tableau is the number of the boxes to the right plus the number of boxes below, plus one, equal to the numbers written in the following example diagram λ :

8	6	5	4	2	1
5	3	2	1		
1					

. For this diagram and $N = 10$ the right side is $2 \ln(5, 945, 940)$.

Example (minimal model of type $(p + 1, p)$ in 1+1 dimensions) The irreducible inequivalent representations are labeled by a pair (m, n) of natural numbers. It is discussed in [52] how these representations can be implemented by localized endomorphisms. The statistical dimensions of the corresponding endomorphisms are

$$\dim(\rho_{(m,n)}) = (-1)^{n+m} \frac{\sin\left(\frac{\pi(p+1)m}{p}\right) \sin\left(\frac{\pi pn}{p+1}\right)}{\sin\left(\frac{\pi(p+1)}{p}\right) \sin\left(\frac{\pi p}{p+1}\right)}. \tag{4.148}$$

It is interesting that a similar bound as in Proposition 8 can be obtained for the entanglement measure E_M defined in Sect. 3.6. To set things up, we consider the vacuum representation of the quantum field theory. The vacuum vector $|0\rangle$ is cyclic and separating for $\mathfrak{A}_A \vee \mathfrak{A}_B$ by the Reeh-Schlieder theorem and therefore defines a natural cone $\mathcal{P}^\sharp$. Any state of the QFT ω with finite energy has a vector representative in $\mathcal{P}^\sharp$ that is cyclic for $\mathfrak{A}_A$, $\mathfrak{A}_B$ and for $\mathfrak{A}_E = (\mathfrak{A}_A \vee \mathfrak{A}_B)'$, again by the Reeh-Schlieder theorem. It follows that for such states, the standing assumption made in Sect. 3.6 holds. We now consider a state ω with finite energy and a localized endomorphism ρ such that $\rho^* \omega$ has finite energy.

114 4 Upper Bounds for E_R in QFT

Proposition 9 *Under the same hypothesis as in Proposition 8, if $d + 1 > 2$ we have*

$$0 \leq E_M(\omega) - E_M(\rho^*\omega) \leq \ln \prod_i \dim(\rho_i)^{5n_i/2}. \tag{4.149}$$

Proof Consider first an irreducible ρ. In $d + 1 > 2$ dimensions the DR reconstruction theorem applies and the left-inverse of ρ [see Eq. (4.137)] is given by $\Psi_\rho(x) = N^{-1} \sum_i \psi_i^* x \psi_i$ where $N = \dim(\rho) \in \mathbb{N}$ and where ψ_i are the generators of the Cuntz algebra (2.24). By Proposition 5, we have $E_M(\rho^*\omega) \leq E_M(\omega)$, which is the first inequality. As in the proof of Proposition 8, we also have $N^2 \Psi_\rho^* \rho^* \omega \geq \omega$ from the Pimsner-Popa inequality. By Proposition 3,

$$E_M(\omega) \leq E_M(\Psi_\rho^* \rho^* \omega) + \ln N. \tag{4.150}$$

Now consider the linear map $\phi_N : \mathfrak{A}_A \vee \mathfrak{A}_B \to M_N(\mathbb{C})^{\otimes 2} \otimes (\mathfrak{A}_A \vee \mathfrak{A}_B)$ defined by

$$\phi_N(x) = \sum_{i,j=1}^N |i\rangle\langle j| \otimes 1_N \otimes \psi_i^* x \psi_j. \tag{4.151}$$

We get $\phi_N(x)^* = \phi_N(x^*)$, and the relations of the Cuntz algebra (2.24) furthermore give $\phi_N(1) = 1_N \otimes 1_N \otimes 1$ and

$$\phi_N(x)\phi_N(y) = \sum_{i,j=1}^N |i\rangle\langle j| \otimes 1_N \otimes \psi_i^* x \left(\sum_{k=1}^N \psi_k \psi_k^* \right) y \psi_j = \phi_N(xy), \tag{4.152}$$

so ϕ_N is a unital *-homomorphism. Next, let $\varphi = \rho^*\omega$ and let ω_N^+ be the maximally entangled state on $M_N(\mathbb{C})^{\otimes 2}$. The definitions imply $(\omega_N^+ \otimes \varphi)(\phi_N(x)) = \varphi(\Psi_\rho(x))$. Proposition 5 now gives $E_M(\phi_N^*(\omega_N^+ \otimes \varphi)) \leq E_M(\omega_N^+ \otimes \varphi)$. On the other hand, the tensor product property (e5) of E_M together with $E_M(\omega_N^+) = \frac{3}{2} \ln N$ gives $E_M(\omega_N^+ \otimes \varphi) \leq E_M(\varphi) + \frac{3}{2} \ln N$. Putting this together gives $E_M(\Psi_\rho^* \rho^* \omega) \leq E_M(\rho^*\omega) + \frac{3}{2} \ln N$. Combining with (4.150), we thus get $E_M(\omega) \leq E_M(\rho^*\omega) + \frac{5}{2} \ln N$, which is the claim of the proposition for irreducible ρ. The general case follows by iterating the argument. $\square$

References

1. G. Lechner, K. Sanders, Modular nuclearity: a generally covariant perspective. Axioms **5**, 5 (2016)
2. K. Sanders, On the reeh-schlieder property in curved spacetime. Commun. Math. Phys. **288**, 271–285 (2009)
3. D. Buchholz, C. D'Antoni, R. Longo, Nuclear maps and modular structures. 1. General properties. J. Funct. Anal. **88**, 223 (1990)

4. D. Buchholz, E.H. Wichmann, Causal independence and the energy level density of states in local quantum field theory. Commun. Math. Phys. **106**, 321 (1986)
5. C. D'Antoni, S. Hollands, Nuclearity, local quasiequivalence and split property for Dirac quantum fields in curved space-time. Commun. Math. Phys. **261**, 133 (2006)
6. K. Fredenhagen, K.H. Rehren, B. Schroer, Superselection sectors with braid group statistics and exchange algebras. Commun. Math. Phys. **125**, 201–226 (1989)
7. K. Fredenhagen, A remark on the cluster theorem. Commun. Math. Phys. **97**, 461–463 (1985)
8. A. Jaffe, High energy behavior in quantum field theory I. Strictly localizable fields. Rev. Phys. **158**, 1454 (1967)
9. A.E. Ingham, A note on fourier transforms. J. Lon. Math. Soc. **S1–9**, 29–32 (1934)
10. C.J. Fewster, L.H. Ford, Probability distributions for quantum stress tensors measured in a finite time interval. Phys. Rev. D **92**, 105008 (2015)
11. H.J. Borchers, R. Schumann, A Nuclearity condition for charged states. Lett. Math. Phys. **23**, 65 (1991)
12. Ch. Jäkel, *Cluster Estimates for Modular Structures*, arXiv:hep-th/9804017
13. D. Buchholz, P. Jacobi, On the nuclearity condition for massless fields. Lett. Math. Phys. **13**, 313 (1987)
14. S. Hollands, O. Islam, K. Sanders, *Relative entanglement entropy for widely separated regions in curved spacetime*. J. Math. Phys. **59**, 062301 (2018)
15. R.M. Wald, *Quantum field theory in curved space-time and black hole thermodynamics* (Chicago University Press, Chicago, 1994)
16. P. Gilkey, *Invariance theory, the heat equation and the Atiyah-Singer Index Theorem* (Publish or Perish, Washington, 1984)
17. J.J. Bisognano, E.H. Wichmann, On the duality condition for quantum fields. J. Math. Phys. **17**, 303 (1976)
18. D. Buchholz, G. Lechner, Modular nuclearity and localization. Ann. Henri Poincaré **5**, 1065 (2004)
19. J.L. Cardy, O.A. Castro-Alvaredo, B. Doyon, Form factors of branch-point twist fields in quantum integrable models and entanglement entropy. J. Statist. Phys. **130**, 129 (2008)
20. S. Alazzawi, G. Lechner, *Inverse scattering and locality in integrable quantum field theories*, arXiv:1608.02359 [math-ph]
21. G. Lechner, Construction of quantum field theories with factorizing S-matrices. Commun. Math. Phys. **277**, 821 (2008)
22. M. Kardar, *Statistical physics of particles* (U Press, Cambridge, 2007)
23. R. Brunetti, D. Guido, R. Longo, Modular structure and duality in conformal quantum field theory. Commun. Math. Phys. **156**, 201 (1993)
24. G. Mack, All unitary ray representations of the conformal group SU(2,2) with positive energy. Commun. Math. Phys. **55**, 1 (1977)
25. D. Buchholz, C. D'Antoni, R. Longo, Nuclearity and thermal states in conformal field theory. Commun. Math. Phys. **270**, 267–293 (2007)
26. P.D. Hislop, R. Longo, Modular structure of the local algebras associated with the free massless scalar field theory. Commun. Math. Phys. **84**, 71 (1982)
27. F. Hansen, The fast track to Löwner's theorem. Lin. Alg. Appl. **438**, 4557–4571 (2013)
28. E. Nelson, Analytic vectors, Ann. Math. **70**, 572–615 (1959); R. Goodman, Analytic and entire vectors for representations of Lie groups, Trans. Amer. Mat. Soc. **143**, 55–76 (1969)
29. K. Fredenhagen, J. Hertel, Local algebras of observables and point-like localized fields. Commun. Math. Phys. **80**, 555 (1981)
30. G. Mack, A. Salam, Finite component field representations of the conformal group. Ann. Phys. **53**, 174 (1969)
31. H. Bostelmann, Phase space properties and the short distance structure in quantum field theory. J. Math. Phys. **4**, 052301 (2005)
32. L. Hörmander, *The analysis of linear partial differential operators I* (Springer, Berlin, Heidelberg, 1990)
33. R.M. Wald, *General relativity* (University of Chicago Press, Chicago, 1984)

34. P. DiFrancesco, P. Mathieu, D. Senechal, *Conformal field theory* (Springer, New York, 1997)
35. Y. Kawahigashi, R. Longo, M. Müger, Multi-Interval subfactors and modularity of representations in conformal field theory. Commun. Math. Phys. **219**, 631 (2001)
36. P. Calabrese, J. Cardy, Entanglement entropy and conformal field theory. J. Phys. A **42**, 504005 (2009)
37. R. Longo, F. Xu, *Relative Entropy in CFT*, arXiv:1712.07283 [math.OA]
38. H. Casini, M. Huerta, Remarks on the entanglement entropy for disconnected regions. JHEP **0903**, 048 (2009)
39. S. Doplicher, R. Haag, J.E. Roberts, Local observables and particle statistics. 1. Commun. Math. Phys. **23**, 199 (1971)
40. S. Doplicher, R. Haag, J.E. Roberts, Local observables and particle statistics. 2. Commun. Math. Phys. **35**, 49 (1974)
41. S. Doplicher, R. Longo, Standard and split inclusions of von Neumann algebras. Invent. Math. **75**, 493 (1984)
42. R. Longo, Index of subfactors and statistics of quantum fields. I. Commun. Math. Phys. **126**, 217 (1989)
43. R. Longo, Index of subfactors and statistics of quantum fields. 2: Correspondences, braid group statistics and Jones polynomial. Commun. Math. Phys. **130**, 285 (1990)
44. R. Haag, *Local quantum physics: Fields, particles, algebras* (Springer, Berlin, 1992)
45. K. Fredenhagen, Superselection sectors, Notes from lectures at Hamburg University (1994/1995)
46. H. Araki, *Mathematical theory of quantum fields* (Oxford Science Publications, 1993)
47. S. Doplicher, J.E. Roberts, Why there is a field algebra with a compact gauge group describing the superselection structure in particle physics. Commun. Math. Phys. **131**, 51 (1990)
48. P. Caputa, M. Nozaki, T. Takayanagi, Entanglement of local operators in large-N conformal field theories. Prog. Theor. Exp. Phys. **2014**, 093B06 (2014)
49. M. Nozaki, T. Numasawa, T. Takayanagi, Quantum entanglement of local operators in conformal field theories. Phys. Rev. Lett. **112**, 111602 (2014)
50. M. Pimsner, S. Popa, Entropy and index for subfactors. Ann. Sci. Ecole Norm. Sup. **19**, 57 (1986)
51. M. Ohya, D. Petz, *Quantum entropy and its use, Theoretical and Mathematical Physics* (Springer, Berlin, Heidelberg, 1993)
52. Y. Kawahigashi, R. Longo, Classification of local conformal nets: Case c < 1. Ann. Math. **160**, 493 (2004)

Chapter 5
Lower Bounds

Abstract In this chapter we derive some lower bounds for the relative entanglement entropy. We include lower bounds of area law type for ground states of suitable QFTs and some general lower bounds for generic states.

5.1 Lower Bounds of Area Law Type

As one may guess from the definition of E_R (infimum over separable comparison states), it is not evident how to obtain lower bounds. In fact, it is not even entirely obvious that $E_R(\omega) > 0$, say, in the vacuum. We first settle this question.

Corollary 1 *Let ω be any state such that the conclusions of the Reeh-Schlieder theorem hold, such as the vacuum ω_0, a KMS-state ω_β, or any state with bounded energy in a Minkowski quantum field theory. Let A, B be open non-empty regions with* dist$(A, B) > 0$. *Then $E_R(\omega) > 0$.*

Proof Our proof is rather similar to that of [1], which in turn is based on the works of [2, 3]. As usual, we represent our net $\{\mathfrak{A}(O)\}$ on a Hilbert space via the GNS-construction, which gives a representation π on $\mathcal{H}$ such that ω is represented by a vector $|\Omega\rangle$. We write $\mathfrak{A}_A = \pi(\mathfrak{A}(O_A))''$, where O_A is the causal diamond with base A, and similarly for B.

Assume that $E_R(\omega) = 0$. By Corollary 2 below there exists, for each $\delta > 0$, a separable state $\omega' = \sum_j \varphi_j \otimes \psi_j$ with positive normal functionals φ_j, ψ_j such that $\|\omega - \omega'\| < \delta$. Using the split property, one can choose a type I subalgebra $\mathfrak{W}_A$ of the type III$_1$-algebra $\mathfrak{A}_A$. This subalgebra may be chosen to be a factor on some Hilbert space $\mathcal{H}_A$ and may be realized as the v. Neumann closure of a Weyl-algebra for one degree of freedom ("Cbit"), i.e. we may think of $\mathfrak{W}_A$ as being isomorphic to $\mathfrak{W}_A \cong \mathfrak{W}(\mathbb{R}^2, \sigma_2)''$, where $\sigma_2 = \begin{pmatrix} 0 & 1 \\ -1 & 0 \end{pmatrix}$ is the standard symplectic form on $\mathbb{R}^2$. The

© The Author(s), under exclusive licence to Springer Nature Switzerland AG 2018   117
S. Hollands and K. Sanders, *Entanglement Measures and Their Properties
in Quantum Field Theory*, SpringerBriefs in Mathematical Physics 34,
https://doi.org/10.1007/978-3-319-94902-4_5

same construction can of course be made for B. We now choose a state η on $\mathfrak{W}_A \otimes$ $\mathfrak{W}_B \cong \mathfrak{W}(\mathbb{R}^2 \oplus \mathbb{R}^2, \sigma_2 \oplus \sigma_2)''$ such that $E_B(\eta)$ (the entanglement measure defined in Sect. 3.3) satisfies $E_B(\eta) \geq \sqrt{2} - \epsilon$ for some small $\epsilon > 0$, i.e. a state in which the Bell-inequality is nearly maximally violated. Extend η to a state on $\mathfrak{A}_A \vee \mathfrak{A}_B$ via the Hahn-Banach theorem.[1] The extended state, called again η, need not be normal to ω. But by Fell's theorem (see e.g. [4]), we can choose a normal state ψ which approximates η arbitrarily well in the weak topology on the subalgebra $\mathfrak{W}_A \otimes \mathfrak{W}_B$. In particular, by choosing suitable operators in Eq. (88) we can achieve that $E_B(\psi) \geq E_B(\eta) - \epsilon \geq \sqrt{2} - 2\epsilon$ for arbitrarily small ϵ. Let $|\Psi\rangle \in \mathcal{H}$ be the unique representer of this ψ in the natural cone of $|\Omega\rangle \in \mathcal{H}$ (the GNS-representative of ω). Since we are assuming the Reeh-Schlieder property, $|\Omega\rangle$ is both cyclic and separating for $\mathfrak{A}_A$, say, and so we can find an a from this algebra such that $a|\Omega\rangle$ approximates $|\Psi\rangle$ arbitrarily well and such that $\||a|\Omega\rangle\| = 1$. Let $\varphi = \omega(a^* . a)$ be the corresponding positive functional on $\mathfrak{A}_A \vee \mathfrak{A}_B$. In particular, we may choose a such that $E_B(\varphi) \geq \sqrt{2} - 3\epsilon$. Next, consider $\varphi' = \omega'(a^* . a) = \sum_j \varphi_j(a^* . a) \otimes \psi_j$. Clearly, φ' is separable (so $E_B(\varphi') = 1$), and

$$\|\varphi - \varphi'\| \leq \|a\|^2 \|\omega - \omega'\| < \delta\|a\|^2 . \tag{5.1}$$

Therefore, by choosing δ sufficiently small, we can achieve that $E_B(\varphi) \leq 1 + \epsilon$ (the invariant E_B is norm continuous). This is in contradiction with $E_B(\varphi) \geq \sqrt{2} - 3\epsilon$ for sufficiently small ϵ. $\square$

The lower bound we have just derived is of course not satisfactory and only serves to confirm our expectation that the invariant E_R is non-trivial in the context of quantum field theory. To get $E_R(\omega) > 0$ in the previous proof, we employed a pair of type I subalgebras $\mathfrak{W}_A \subset \mathfrak{A}_A$, $\mathfrak{W}_B \subset \mathfrak{A}_B$, each isomorphic to the algebra of one continuous quantum mechanical degree of freedom ("Cbit"). We showed that for a large class of states such as the vacuum ω_0, the restriction to $\mathfrak{W}_A \otimes \mathfrak{W}_B$, i.e. our Cbit pair, is entangled.

To obtain a better lower bound, we now pass to a large number N of Cbits embedded into disjoint subregions $A_i \subset A$ and $B_i \subset B$, where $i = 1, \ldots, N$. The idea is that each of these N copies will contribute at least one Cbits' worth of entanglement, and thus give us a much better lower bound. This will work as stated if the entanglement measure E satisfies the strong superadditivity property (e6). In this situation, we thus expect an entanglement at least proportional to N (because each of the N Cbit pairs is expected to contribute one unit), while N itself is restricted only by the requirement that the regions A_i, B_i should be non-intersecting, i.e. a geometrical property. For an entanglement measure E that does not fulfill (e6)—like for instance E_R—we will argue via an auxiliary measure—like E_D—which does.

Our reasoning will work most straightforwardly for a conformal field theory in $d+1$ dimensions, and for simplicity we will stick to these theories here. We are

---

[1]This theorem gives a bounded extension $\hat{\eta}$ with norm $\|\hat{\eta}\| \leq \|\eta\| = \eta(1)$. Since $1 \in \mathfrak{W}_A \otimes \mathfrak{W}_B$, we have $\hat{\eta}(1) = 1$, and we may also take $\hat{\eta}$ to be hermitian. If not, we take instead $\Re\hat{\eta}$. It follows that $\Re\hat{\eta}$ is also positive, i.e. a state.

interested primarily in the case when there is only a "thin corridor" of size ϵ between A and B. To formalize this, we take A to have a smooth boundary ∂A and outward unit normal n. We can "flow" the boundary outwards along the geodesics tangent to n by a small proper distance $\epsilon > 0$. In this way, we obtain a slightly larger region $A_\epsilon \supset A$, and we let $B \subset \mathbb{R}^d \backslash \overline{A}_\epsilon$. The proof of the following simple theorem was inspired by conversations with J. Eisert [5].

Theorem 17 *Given a net in spatial dimensions $d \geq 1$ with invariance group consisting of Poincaré-transformations and dilations, and with invariant vacuum state ω_0, and given regions A, B separated by a thin corridor of size ϵ we have for $\epsilon \to 0$*

$$E_R(\omega_0) \gtrsim \begin{cases} D_2 \cdot \frac{|\partial A|}{\epsilon^{d-1}} & \text{when } d \geq 2, \\ D_2' \cdot \ln \frac{\min(|A|,|B|)}{\epsilon} & \text{when } d = 1, \end{cases} \quad (5.2)$$

where $|\partial A|$ is the surface area of the boundary when $d \geq 2$, where $|A|$, $|B|$ denote the lengths of the intervals when $d = 1$, where D_2 is the distillable entropy of one Cbit pair (defined more precisely in the proof) and $D_2' = D_2 \log_3 e$.

Remark 1 Instead of E_R, one can obtain the same result obviously for any other entanglement measure dominating E_D which satisfies (e4), or with any entanglement measure obeying (e4) for automorphisms and (e6), such as an appropriate generalization of the "squashed entanglement" E_S [6] for type III factors.

Proof ($d \geq 2$): We consider a pair consisting of a unit cube $A_0 = c = (0, 1)^d$ at the origin in a spatial slice $\cong \mathbb{R}^d$, and a unit cube B_0 obtained from A_0 by a translation in some arbitrarily chosen coordinate direction. We fix the distance between A_0 and B_0 to be, say, one. As in the previous proof, we embed a Cbit pair (i.e. pair of type I algebras $\mathfrak{W}_{A_0}$, $\mathfrak{W}_{B_0}$ each isomorphic to the v. Neumann closure of the Weyl algebra of one continuous quantum mechanical degree of freedom) into $\mathfrak{A}_{A_0}$, $\mathfrak{A}_{B_0}$, respectively, and call D_2 the distillable entropy of this pair in the restriction of the state ω_0.

We can apply to this pair (A_0, B_0) group elements $\{g_i\}$ of the invariance group generated by dilations, rotations, and spatial translations so that each $(A_i, B_i) = g_i \cdot (A_0, B_0)$ is a pair of cubes of size 2ϵ lying on opposite sides of the corridor separating A from B, see Fig. 5.1. We assume that $1/\epsilon$ is much larger than the maximum of the extrinsic curvature $(K_{ij} K^{ij})^{1/2}$ along ∂A, so that the boundary is essentially flat on the scale ϵ. If we demand that the cube pairs do not intersect with each other, then it is clear that we can fit in $N \gtrsim |\partial A|/\epsilon^{d-1}$ cube pairs (asymptotically for $\epsilon \to 0$). Defining $\mathfrak{W}_{A_i} = \alpha_{g_i} \mathfrak{W}_{A_0}$ (and similarly for B_i), we then have an inclusion $\iota_N : \otimes_i \mathfrak{W}_{A_i} \to \mathfrak{A}_A$ (and similarly for B). Let ω_i be the restriction of ω to $\mathfrak{W}_{A_i} \otimes \mathfrak{W}_{B_i}$ under this inclusion. The properties of E_R, E_D imply:

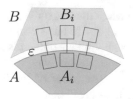

Fig. 5.1 The the sets A_i, B_i in $d + 1 > 2$ spacetime dimensions

$$
E_R(\omega_0) \geq E_R\left(\omega_0 \upharpoonright \bigvee_{i=1}^{N} \mathfrak{A}_{A_i} \otimes \mathfrak{A}_{B_i}\right)
$$

$$
\geq E_R\left(\omega_0 \upharpoonright \bigvee_{i=1}^{N} \mathfrak{W}_{A_i} \otimes \mathfrak{W}_{B_i}\right)
$$

$$
\geq E_D\left(\omega_0 \upharpoonright \bigvee_{i=1}^{N} \mathfrak{W}_{A_i} \otimes \mathfrak{W}_{B_i}\right) \tag{5.3}
$$

$$
\geq \sum_{i=1}^{N} E_D(\omega_0 \upharpoonright \mathfrak{W}_{A_i} \otimes \mathfrak{W}_{B_i})
$$

$$
= \sum_{i=1}^{N} E_D(\alpha_{g_i}^* \omega_0 \upharpoonright \mathfrak{W}_{A_0} \otimes \mathfrak{W}_{B_0}) = N \cdot D_2 \gtrsim D_2 \cdot \frac{|\partial A|}{\epsilon^{d-1}}.
$$

In the first and second step, we used (e4) (letting $\mathcal{F}$ be the inclusion map $\vee \mathfrak{A}_{A_i} \otimes \vee \mathfrak{A}_{B_i} \to \mathfrak{A}_A \otimes \mathfrak{A}_B$ in the first, and $\vee \mathfrak{W}_{A_i} \otimes \vee \mathfrak{W}_{B_i} \to \vee \mathfrak{A}_{A_i} \otimes \vee \mathfrak{A}_{B_i}$ the second step). In the third step we used that E_R dominates E_D for type I algebras, by Theorem 5. In the fourth step we used (e6) for E_D. In the fifth step we used that ω_0 is invariant under α_{g_i} (conformal invariance of the vacuum), and that $E_D(\omega_0 \upharpoonright \mathfrak{W}_{A_0} \otimes \mathfrak{W}_{B_0}) = D_2$ by definition. ($d = 1$): By dilation invariance, we may assume without loss of generality that $\min(|A|, |B|) = 1$. A cube is now an interval, and we consider the interval pairs

$$
A_i = (-(\tfrac{1}{3})^i, -(\tfrac{1}{3})^{i+1}), \quad B_i = ((\tfrac{1}{3})^{i+1}, (\tfrac{1}{3})^i), \tag{5.4}
$$

see Fig. 5.2. These intervals are obviously disjoint and they satisfy $A_i \subset A$ respectively $B_i \subset B$ as long as $i + 1 \leq \lfloor \log_3 \epsilon^{-1} \rfloor$. The number N of (A_i, B_i)-pairs is thus $\sim \log_3 \epsilon^{-1}$ when $\epsilon \to 0$. The rest of the proof then follows the same argument as in the case $d \geq 2$. □

That generic states satisfying the Reeh-Schlieder property are distillable across a pair of spacelike regions has been shown in a rather general setting by [7]. Here, we would like to ensure that the distillation *rate* for the vacuum state is in fact non-zero—or more precisely that $D_2 > 0$—which is a stronger statement. We now present an argument that this must be the case at least for the massless free KG field which defines a conformal net: Since this theory satisfies the Reeh-Schlieder property, we can argue

Fig. 5.2 The sets A_i, B_i in $d + 1 = 2$ spacetime dimensions

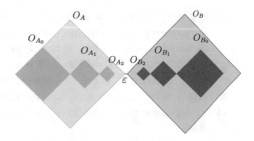

just as in the proof of Corollary 1 that ω restricted to $\mathfrak{W}_{A_0} \vee \mathfrak{W}_{B_0}$ cannot be separable. Since this algebra is isomorphic to the weak closure of $\mathfrak{W}(\mathbb{R}^2, \sigma_2) \otimes \mathfrak{W}(\mathbb{R}^2, \sigma_2) \cong \mathfrak{W}(\mathbb{R}^4, \sigma_4)$ in the restriction of the vacuum representation to this subalgebra, and since the vacuum state is quasi-free, its restriction to $\mathfrak{W}(\mathbb{R}^4, \sigma_4)$ must also be a quasi-free state. For such states it is known [8–10] that they cannot have a property called "positive partial transpose". Using this it is shown in [11] that such states satisfy a "reduction criterion" which in [12] was shown to imply a finite distillable entropy. Hence $D_2 > 0$ for the free massless KG field.[2]

Remark 2 Looking at the proof, one sees that one could replace the elementary Cbit pair by N continuous quantum mechanical degrees of freedom, i.e. by replacing $\mathfrak{W}_{A_0}$ resp. $\mathfrak{W}_{B_0}$ with a v. Neumann algebra isomorphic to the weak closure of $\mathfrak{W}(\mathbb{R}^{2N}, \sigma_{2N})$ sitting inside $\mathfrak{A}_{A_0}$ resp. $\mathfrak{A}_{B_0}$, and then defining $E_D(\omega_0 \restriction \mathfrak{W}_{A_0} \otimes \mathfrak{W}_{B_0}) = D_N$. We can then maximize over the parameter N, and in the lower bound (5.2) D_2 is then replaced by the maximum possible D_N. For a free scalar field with N components, this yields an improvement of the lower bound by the factor N since $D_N = ND_2$. More generally, the conclusions of the theorem are likely to be true for any state ω that is asymptotically dilation invariant on small scales (e.g. states with finite energy) and for any theory which approaches a free field theory on small scales, i.e. any asymptotically free theory. Thus, it is highly plausible that the following bound holds for an asymptotically free theory and any state with finite energy:

$$E_R(\omega) \gtrsim N \cdot D_2 \cdot \frac{|\partial A|}{\epsilon^{d-1}}, \tag{5.5}$$

where N is the number of independent free fields in the scaling limit.

5.2 General Lower Bounds

One can use the variational definition of E_R to obtain some (rather indirect) lower bounds in terms of the norm distance of ω to the subspace of separable states. We now explain these—essentially well-known—bounds. Returning to the general situation,

---

[2]See also [13] for further discussion on the distillation of quasi-free states.

let $\mathfrak{A}$ be a v. Neumann algebra, and ω, ω' two faithful normal states. Then $\varphi(a) = \omega(a) - \omega'(a)$ is a linear, hermitian, continuous, non-positive functional on $\mathfrak{A}$. For any such functional, one can define its "range projection", $e = e(\varphi) \in \mathfrak{A}$.[3] For instance, if φ is the functional defined by $\varphi(a) = \mathrm{Tr}(Fa)$ on a matrix v. Neumann algebra $\mathfrak{A} = M_n(\mathbb{C})$ in terms of some self-adjoint matrix $F = F^*$, the range projection e would be given by the projection onto the non-negative eigenvalues of F. It follows from this definition that the norm of any hermitian linear functional on $\mathfrak{A}$ is given in this case by

$$\|\varphi\| = \sup_{a \in \mathfrak{A}, \|a\| \leq 1} |\varphi(a)| = \varphi(e) - \varphi(1 - e), \tag{5.6}$$

as one can easily prove by showing that the left side is neither bigger nor smaller than the right side. These formulas generalize to general continuous, hermitian, linear functionals on arbitrary v. Neumann algebras $\mathfrak{A}$, and one can show that $e = e^* = e^2$ is always an element of $\mathfrak{A}$. For the norm of $\omega - \omega'$ we therefore get

$$\|\omega - \omega'\| = (\omega - \omega')(e) - (\omega - \omega')(1 - e) = 2(p - q), \tag{5.7}$$

where $0 < q \leq p < 1$ have been defined by $p = \omega(e)$, $q = \omega'(e)$. Consider now the subalgebra $\mathfrak{D}$ of $\mathfrak{A}$ generated by $\{e, 1\}$. It is obvious that this subalgebra is abelian and isomorphic to the trivial v. Neumann algebra of diagonal complex 2 by 2 matrices, and under this isomorphism, the restrictions $\omega|_{\mathfrak{D}}, \omega'|_{\mathfrak{D}}$ correspond to the diagonal density matrices

$$\rho_{\mathfrak{D}} = \begin{pmatrix} p & 0 \\ 0 & 1 - p \end{pmatrix}, \quad \rho'_{\mathfrak{D}} = \begin{pmatrix} q & 0 \\ 0 & 1 - q \end{pmatrix}. \tag{5.8}$$

Combining this with (h2) (for the inclusion map $\mathcal{F} : \mathfrak{D} \to \mathfrak{A}$), we get

$$H(\omega, \omega') \geq H(\omega|_{\mathfrak{D}}, \omega'|_{\mathfrak{D}}) = p \ln \frac{p}{q} + (1 - p) \ln \frac{1 - p}{1 - q}. \tag{5.9}$$

Hiai et al. [15] estimate the right side as $\geq 2(p - q)^2$, which in view of (5.7) immediately gives the well-known result, also stated by these authors,

$$H(\omega, \omega') \geq \tfrac{1}{2} \|\omega - \omega'\|^2. \tag{5.10}$$

By a trivial modification of the argument, one can obtain a tighter lower bound. Define $s(x)$ to be the infimum of the right hand side of (5.9) under the constraint $p - q = x \geq 0$, i.e.

$$s(x) \equiv \inf_{p,q : p - q = x, 0 < q \leq p < 1} \left[p \ln \frac{p}{q} + (1 - p) \ln \frac{1 - p}{1 - q} \right]. \tag{5.11}$$

---

[3]This follows by applying the arguments in the proofs of Theorems 7.3.1 and 7.3.2 in [14] to the self-adjoint part of the unit ball.

In view of $x = \frac{1}{2}\|\omega - \omega'\|$, (5.9) actually gives the improved lower bound

$$H(\omega, \omega') \geq s\left(\tfrac{1}{2}\|\omega - \omega'\|\right). \tag{5.12}$$

The function $s : (0, 1) \to \mathbb{R}$ is monotonically increasing, strictly convex, positive, and has the asymptotic behavior [16]

$$s(x) \sim \begin{cases} 2x^2 + \frac{4}{9}x^4 + \frac{32}{135}x^6 + \ldots & \text{for } x \to 0, \\ -\ln(1 - x) & \text{for } x \to 1. \end{cases} \tag{5.13}$$

From the second line it is seen that the improvement of the lower bound is most drastic when $x \to 1$, i.e. when $\|\omega - \omega'\| \to 2$ (note that 2 is the maximum value since ω, ω' are functionals of norm one). For matrix algebras $\mathfrak{A} = M_N(\mathbb{C})$, where the states ω, ω' can be identified with density matrices $\rho_\omega, \rho_{\omega'}$, the norm distance is $\|\omega - \omega'\| = \|\rho_\omega - \rho_{\omega'}\|_1$, the 1-norm of an operator being defined by Definition 5. Our inequality (5.12) thereby reduces to an inequality found by [16] using a more involved method.

As an aside we note that instead of using the norm $\|\omega - \omega'\|$, one can also obtain a lower bound directly in terms of suitable vector representatives $|\Omega\rangle, |\Omega'\rangle$ in the GNS representation of, say, ω, using Proposition 2. Using also the monotonicity of s, we immediately arrive at:

Theorem 18 *Let ω, ω' be faithful normal states on a v. Neumann algebra $\mathfrak{A}$, with vector representatives $|\Omega\rangle, |\Omega'\rangle \in \mathcal{P}^\sharp$ in the natural cone, so that $1 \geq \langle \Omega' | \Omega \rangle > 0$. Then we have*

$$H(\omega, \omega') \geq s\left(1 - \langle \Omega' | \Omega \rangle\right), \tag{5.14}$$

where $s : (0, 1) \to \mathbb{R}$ is the universal positive monotonic function defined by (5.11).

This lower bound is useful in the context of Gaussian states for free fields, as $\langle \Omega | \Omega' \rangle$ can be expressed in terms of the operators Σ, Σ' defining these states.

Returning from these general considerations to quantum field theory, consider a local net $O \mapsto \mathfrak{A}(O)$, and let O_A and O_B be two causal diamonds with disjoint bases A and B on some Cauchy surface $\mathcal{C}$. As in the description of the split construction above, we assume that there is a safety distance $\text{dist}(A, B) > 0$ between the two bases. Let ω be a faithful normal state on the algebra $\mathfrak{A}_A \vee \mathfrak{A}_B$ (e.g. the vacuum state for the entire net). Then the decoupled state $\omega'(ab) = (\omega \otimes \omega)(ab) := \omega(a)\omega(b)$ is well-defined by the split property. Obviously

$$\|\omega - \omega'\| \geq \frac{(\omega - \omega')(ab)}{\|ab\|} \geq \frac{\omega(ab) - \omega(a)\omega(b)}{\|a\| \cdot \|b\|} \tag{5.15}$$

From the definitions of the mutual information and entanglement entropy of the pair A, B and the monotonicity of s, we immediately get

Corollary 2 *Let O_A and O_B be causal diamonds with bases A and B on some Cauchy surface such that* $\mathrm{dist}(A, B) > 0$. *Then*

$$E_I(\omega) \geq \sup s\left(\frac{\omega(ab) - \omega(a)\omega(b)}{2\|a\| \cdot \|b\|}\right), \qquad (5.16)$$

the supremum being over all nonzero $a \in \mathfrak{A}_A$, $b \in \mathfrak{A}_B$. Similarly

$$E_R(\omega) \geq \inf_{\sigma} s\left(\tfrac{1}{2}\|\omega - \sigma\|\right) \qquad (5.17)$$

where the infimum is over all separable states on $\mathfrak{A}_A \vee \mathfrak{A}_B$.

It is possible to see form the second inequality (5.17) and the asymptotic behavior of s that, if $B = \mathbb{R}^d \backslash A_\epsilon$ as above, then $E_R(\omega)$ must diverge as $\epsilon \to 0$, for any normal state ω.

References

1. H. Narnhofer, Entanglement, split, and nuclearity in quantum field theory. Rep. Math. Phys. **50**, 111 (2002)
2. S.J. Summers, R. Werner, Maximal violation of Bell's inequalities for algebras of observables in tangent spacetime regions. Ann. Inst. H. Poincaré **49**, 2 (1988)
3. S.J. Summers, R. Werner, Maximal violation of Bell's inequalities is generic in quantum field theory. Commun. Math. Phys. **110**, 247–259 (1987)
4. R. Haag, *Local Quantum Physics: Fields, Particles, Algebras* (Springer, Berlin, 1992)
5. J. Eisert, Private Communication
6. M. Christiandl, A. Winter, Squashed entanglement'—an additive entanglement measure. J. Math. Phys. **45**, 829–840 (2004)
7. R. Verch, R.F. Werner, Distillability and positivity of partial transposes in general quantum field systems. Rev. Math. Phys. **17**, 545 (2005)
8. R.F. Werner, M.M. Wolf, Bound entangled Gaussian states. Phys. Rev. Lett. **86**, 3658 (2001)
9. L.-M. Duan, G. Giedke, J.I. Cirac, P. Zoller, Inseparability criterion for continuous variable systems. Phys. Rev. Lett. **84**, 2722 (2000)
10. R. Simon, Peres-Horodecki separability criterion for continuous variable systems. Phys. Rev. Lett. **84**, 2726 (2000)
11. G. Giedke, L.-M. Duan, J.I. Cirac, P. Zoller, Distillability criterion for all bipartite Gaussian states. Quant. Inf. Comput. **1**(3), 79–86 (2001)
12. M. Horodecki, P. Horodecki, Reduction criterion of separability and limits for a class of distillation protocols. Phys. Rev. A **59**, 4206 (1999)
13. J. Eisert, S. Scheel, M.B. Plenio, Distilling Gaussian states with Gaussian operations is impossible. Phys. Rev. Lett. **89**, 13 (2002)
14. R.V. Kadison, J.R. Ringrose, *Fundamentals of the Theory of Operator Algebras* (Academic Press, New York, I 1983, II 1986)
15. F. Hiai, M. Ohya, M. Tsukada, Sufficiency, KMS condition and relative entropy in von Neumann algebras. Pacific J. Math. **96**, 99–109 (1981)
16. K.M.R. Audenaert, J. Eisert, Continuity bounds on the quantum relative entropy. J. Math. Phys. **46**, 102104 (2005)

Appendix

A.1 The Edge of the Wedge Theorem

In the body of the volume, we used several times the edge-of-the-wedge theorem. For the convenience of the reader, we give a statement of this theorem and make some remarks. In its most basic form, the theorem deals with the following situation. $U = (x_1, x_2)$ is an open interval in $\mathbb{R}$, F_1 a function that is holomorphic on the upper half plane of $\mathbb{C}$, F_2 a function holomorphic on the lower half plane, both F_1 and F_2 have the same bounded, continuous limit on U. Then there exists a function F, holomorphic in the cut plane $\mathbb{C}\backslash[(-\infty, x_1] \cup [x_2, \infty)]$, which is a joint extension of F_1, F_2.

A more general version of the theorem applies to analytic functions F_1, F_2 holomorphic on a domain of $\mathbb{C}^n$ of the form $U + iC$ resp. $U - iC$, where $U \subset \mathbb{R}^n$ is an open domain, and where $C \subset \mathbb{R}^n$ is the intersection of some open, convex cone with an open ball. It is assumed that

$$T_1(f) = \lim_{y \in C, y \to 0} \int d^n x \, F_1(x + iy) f(x), \quad T_2(f) = \lim_{y \in C, y \to 0} \int d^n x \, F_2(x - iy) f(x)$$

(A.1)

define distributions on U such that, actually, $T_1 = T_2$. The edge of the wedge theorem is (see e.g. [1]):

Theorem 19 *There exists a function F which is holomorphic on an open complex neighborhood $N \subset \mathbb{C}^n$ containing U such that F extends both F_1, F_2 where defined.*

One often applies the theorem to the case when a holomorphic function F_1 on $U + iC$ is given with distributional boundary value $T_1 = 0$. Then choosing $F_2 \equiv 0$, one learns that also $F_1 = 0$ where defined.

The edge of the wedge theorem has a straightforward generalization to the case when F_1, F_2 take values in a Banach space, $\mathcal{X}$, which we also use in this volume. A function F valued in $\mathcal{X}$ is called (weakly) holomorphic near z_0 if $\psi(F(z))$ is holomor-

© The Author(s), under exclusive licence to Springer Nature Switzerland AG 2018
S. Hollands and K. Sanders, *Entanglement Measures and Their Properties*
in Quantum Field Theory, SpringerBriefs in Mathematical Physics 34,
https://doi.org/10.1007/978-3-319-94902-4

phic near z_0 for any linear functional ψ in the topological dual $\mathcal{X}^*$. It is easy to see (see e.g. [2]) that a weakly holomorphic function is in fact even strongly holomorphic in the sense that it has a norm-convergent expansion $F(z) = \sum_{n\geq0} x_n(z - z_0)^n$, $x_n \in \mathcal{X}$ near z_0 (and of course vice versa). By going through the proof of the edge of the wedge-theorem in the $\mathbb{C}$-valued case, one can see as a consequence that an $\mathcal{X}$-valued version holds true, too: if F_i are holomorphic $\mathcal{X}$-valued functions in $U \pm iC$ such that their distributional boundary values (A.1) (limit in the norm topology on $\mathcal{X}$) on U coincide as distributions valued in $\mathcal{X}$, then there is a holomorphic extension F on N. For a related discussion, see also [3].

We also use in this volume the following (related) lemma about $\mathcal{X}$-valued holomorphic functions.

Lemma 14 *Let U be an open domain in $\mathbb{C}$ and let $F : U \to \mathcal{X}$ be a holomorphic function with continuous limit on ∂U. Then $\overline{U} \ni z \mapsto \|F(z)\|_{\mathcal{X}}$ assumes its maximum on the boundary ∂U.*

Proof The norm $u(z) \equiv \|F(z)\|_{\mathcal{X}}$ is continuous on $\overline{U}$ and for each continuous linear map $l : \mathcal{X} \to \mathbb{C}$ the scalar function $l \circ F$ is continuous on $\overline{U}$ and holomorphic on the interior. If $\mathcal{X}_1^*$ denotes the unit ball of the dual space $\mathcal{X}^*$, then

$$
\begin{aligned}
\max_{z\in\overline{U}} u(z) &= \max_{z\in\overline{U}} \max_{l\in\mathcal{X}_1^*} |l(F(z))| = \max_{l\in\mathcal{X}_1^*} \max_{z\in\overline{U}} |l(F(z))| \\
&= \max_{l\in\mathcal{X}_1^*} \max_{z\in\partial U} |l(F(z))| = \max_{z\in\partial U} \max_{l\in\mathcal{X}_1^*} |l(F(z))| = \max_{z\in\partial U} u(z),
\end{aligned}
\tag{A.2}
$$

where we applied the maximum principle to $l \circ F$ to get to the second line. □

References

1. R.F. Streater, A.S. Wightman, *PCT, Spin and Statistics, and All That* (Princeton University Press, Princeton, 2000)
2. J. Feldman, Analytic Banach Space Valued Functions, http://www.math.ubc.ca/~feldman/m511/analytic.pdf
3. A. Strohmaier, R. Verch, M. Wollenberg, Microlocal analysis of quantum fields on curved space-times: analytic wavefront sets and Reeh-Schlieder theorems. J. Math. Phys. **43**, 5514 (2002)

Printed in the United States
By Bookmasters